AMERICAN
MUSICAL COMEDY

AMERICAN MUSICAL COMEDY

From *Adonis* to *Dreamgirls*

GERALD BORDMAN

New York   Oxford
OXFORD UNIVERSITY PRESS
1982

Copyright © 1982 by Oxford University Press, Inc.

Library of Congress Cataloging in Publication Data

Bordman, Gerald Martin.
American musical comedy.

Includes index.
1. Musical revue, comedy, etc.—United States.
I. Title.
ML1711.B66   782.81'0973   81-22444
    ISBN 0-19-503104-0   AACR2

Printing (last digit):   9 8 7 6 5 4 3 2 1

Printed in the United States of America

Preface

This is the second volume in a study of specific genres in American musical theatre. The first volume, *American Operetta: From H.M.S. Pinafore to Sweeney Todd*, presented the thesis that the modern American "musical play" has evolved naturally and ineluctably from nineteenth-century comic opera.

A detailed examination of American musical comedy reveals no such clean-cut development, which is in itself an interesting revelation. While operetta and revue went their independent ways, musical comedy borrowed haphazardly from both genres and responded to numerous influences outside the theatre. The changes that occurred, the differences, say, between *Adonis* and *George Washington, Jr.*, or *Funny Face* and *Dreamgirls*, are fascinating in themselves, however, and are historically important to an understanding of American musical theatre.

I am especially grateful for the help given me in preparing this book by the staffs of the British Museum, the Library of Congress, the Library of the Performing Arts (New York City), the Theatre Collection at the Museum of the City of New York, the Theatre Collection of the Free Library of Philadelphia, and the library at Millersville (Pa.) State Teachers College.

Nor can I ignore the invaluable cooperation and advice of my superb copy editor, Kim Lewis; my editor's learned and amazingly productive secretary and my typist, Joellyn Ausanka; and, most of all, my loyal and constructive editor, Sheldon Meyer.

Yellow Wood Farm   G.B.
Kirk's Mills, Pa.
May 1982

Contents

AMERICAN
MUSICAL COMEDY

1

Beginnings

A few years after *Kiss Me, Kate* opened in 1948, its librettists, Bella and Sam Spewack, were asked to differentiate between new and old musical comedy, the spanking new school of musical comedy that *Kiss Me, Kate* represented and the seemingly superannuated school that had flourished until a few years before. "You may remember," they responded, "that the old musical comedy consisted of a story, songs, dances, scenery, girls and boys. On the other hand, the New Art Form consists of a story, songs, dances, scenery, girls and boys." The Spewacks' comparisons, though tongue-in-cheek, were absolutely accurate—as far as they went.

Musical comedy, from the oldest to the latest, has always told something of a story, even when that story was unlikely or flimsy or sometimes got lost in a welter of other attractions. At intervals during a show, the story invariably made way for—indeed, sometimes simply was made for—a parade of lighthearted songs and dances. Story, songs, and dances unfolded amid eye-catching scenery and equally eye-catching costumes (one careless omission by the otherwise thoughtful Spewacks). Girls and boys were the final element, the Spewacks concluded—not men and women. Here, too, they touched on an essential of musical comedy, its celebration of youthfulness. Principal players aside, the girls were more important than the boys. Early on, a long line of beautiful young ladies, who might or might not be able to sing and dance well, became a fixture of musical comedy. In a few instances—the Gaiety girls and Cochran's in England, the Casino girls, Weber and Fields's and Ziegfeld's in America—they were as big an attraction as the stars and songs and stories.

Older style operetta and its modern counterpart, the musical play, also presented stories, songs, dances, scenery, girls and boys. What separated operetta and musical comedy were fundamental approaches, fundamental attitudes. At heart, operetta trafficked in a roseate, earnest romanticism, frequently transporting both its characters and its audiences to far-off, exotic lands and far-off, fondly remembered times. By contrast, musical comedy professed to take a jaundiced, cynical look at everyday and, more frequently than not, very contemporary foibles. Distinctions between genres were occasionally blurred; imaginary boundaries were sometimes trespassed, knowingly or unknowingly. By error or by design, authors, producers, and commentators often fudged definitions.

Moreover, like operetta, musical comedy changed. Along with other arts, it evolved incessantly. Just as *H.M.S. Pinafore*, *The Merry Widow*, *Rose-Marie*, and *Oklahoma!* markedly differed from one another while remaining operettas, so *Adonis*, *A Gaiety Girl*, *Sally*, *Anything Goes*, *Pal Joey*, *Kiss Me, Kate*, and *Gypsy* reflected the tastes and nature of their own times while remaining musical comedies.

Inevitably, in such close quarters as the theatrical arena, the two genres interacted as they changed, affecting each other even as each continued to pursue its unique path. At times the paths were far apart, at times they seemed to run parallel to one another, and, briefly, they came together as one. In the end, however, they went separate ways, and since the mid-twenties the two have rarely been confused.

In one respect, the two evolutionary roads were strikingly different. Except during the last heyday of Viennese operetta, 1907–14, the renaissance that began with *The Merry Widow*, the road operetta took, however bumpy now and then, was reasonably straight and was marked by incontestable milestones. Four operettas were towering beacons that pointed the way ahead. *H.M.S. Pinafore* was probably the most important musical ever presented on English-speaking stages. Its popularity in America signaled the real beginnings of the American musical theatre and opened all our stages, many of which had been determinedly closed against such entertainments, to whole evenings of song and dance. In England it signaled the end for the slapdash burlesques that had reigned along the Strand and it gave the popular lyric stage a new respectability. *The Merry Widow* revived an all but dead interest in operetta, not merely on English-speaking stages but around the world. It almost certainly remains the most performed and beloved of all light musicals. But, as will be mentioned again later, it also fostered a certain confusion between operetta and musical comedy. *Rose-Marie*, the first American-made operetta in this quartet, was the biggest international success of the 1920s and sparked a brief, brilliant revival of traditional operetta in America. *Oklahoma!* (along with *Show Boat* before it) redefined some basic conceptions of operetta and even refused to call itself one, but it nevertheless brought operetta back to the contemporary theatre and initiated the greatest era of totally American operetta.

The development of musical comedy was markedly different. The seven musical comedies mentioned earlier were chosen arbitrarily. Another seven might replace them without anyone lifting an eyebrow. Take the lone foreign work in the list: *A Gaiety Girl*. Although the show failed to play George Edwardes's Gaiety Theatre, it is often regarded as the first modern musical comedy, the first of the great Gaiety musicals that were named after the house where many of them did play. But it was not Edwardes's first musical conceived in the new fashion. *In Town* was. Many historians therefore give *In Town* pride of place, although it too did not open at the Gaiety. What was the difference between *In Town* or *A Gaiety Girl* and, say, *A Trip to Chinatown*, the popular American musical comedy that antedated both the West End works? Quite simply, nothing, for there are no earthshaking landmarks along the route of musical comedy.

Schools of operetta and their history are far easier to define, far more pat, than are any schools of musical-comedy writing. Although quibblers may be quick to cite exceptions, it would be possible to write a miniature survey of operetta in America merely by recounting the history of Gilbert and Sullivan, Franz Lehar, Rudolf Friml, and Rodgers and Hammerstein. These men exemplified the basic styles of their eras. What equivalent is there for musical comedies? Certainly the Gaiety musicals could be called a legitimate school. They defined the English musical comedy of their era. Indeed, they have determined the fate of English musical comedy almost up to the present time. But what American school has there been? Harrigan and Hart? Charles Hoyt? The Princess Theatre shows? Rodgers and Hart? Each in its time was ragingly successful, and each spawned a few imitators, but by and large none dominated its lyric stage nor truly pointed the way to the future.

Let's take a few examples. In speaking of Friml an obvious question arises: What about Sigmund Romberg? Could a stranger to both men's operettas, who heard them without being told who wrote which, correctly assign authorship? Probably not. Many lovers of these older operettas still confuse the two. Even in more recent times casual playgoers might still attribute almost every modern musical play to Rodgers and Hammerstein. It is less easy to confuse George Gershwin with Cole Porter or Porter with early Richard Rodgers. The operettas of Friml and Romberg belong to a closely knit family; those musical comedies that Gershwin or Porter or Rodgers composed in the twenties and thirties are at best first cousins.

There is one regrettable stumbling block both operetta and musical comedy throw in the student's path. Their earliest history, the prototypical operettas and musical comedies, can be recorded only sketchily. Not merely do numerous historical gaps remain—lost programs and few early advertisements, not to mention a lack of actual texts—but early musicals were rarely thought of as belonging to definite genres and thus were rarely called by names that might categorize them clearly. Operetta, comic opera, musical comedy were all terms used to describe early lyric efforts to potential playgoers, along with musical farce, musical comedietta, comic musical, and many more. The same slight musical was often pegged with one name at one time or place and with another somewhere else at some other time. Terms were used with cavalier abandon as catchall lures. So, until American musical comedy begins to take on a more or less recognizable form in the last quarter of the nineteenth century, we can do little but glance at some very primitive pioneers of the genre.

These early examples all attest to one fundamental attribute of musical comedy: namely, that in the minds of both authors and audiences the comedy came first, the music second. From the very start, operetta attracted the more musicianly composers, artists carefully trained in every aspect of their art. As a rule, for example, late-nineteenth- and early-twentieth-century operetta composers usually orchestrated and frequently conducted their own works. Musical-comedy composers, at heart little more than melodists, left orchestration and conducting to others. The great American composers who embellished musical comedy so gloriously during the first third of the twentieth century were a lucky, joyous happenstance. And although some of them had a more than rudimentary musical training, their interest remained solely in composing melodies. They continued to leave orchestration and, except on special, showy occasions, conducting to others.

But had their greatness even as melodists been denied us, musical comedy would probably have flourished as strongly if not, in a sense, as enduringly. Alan Jay Lerner has said that a libretto decides whether or not a musical shall be an immediate hit; its music determines whether it survives. Granting that other factors influence a show's success (its performers, its mounting) and that Lerner's rule of thumb applies to operetta as well as musical comedy, comedy has provided musical comedy's basic building blocks, music its decoration. An operetta may succeed even with a stodgy book if its music is soaringly beautiful, but a musical comedy had better provide the laughs its audiences want. That requisite comedy can be and most often tended to be topical. A great clown in an operetta sometimes injected topical allusions and got away with it. But what with their far-off, romantic settings, most operettas confined themselves to more generalized comedy, humor that derived from the situations on stage or from a broader view of human failings. Musical comedy employed the same sort of fun but leaned far more heavily on spoofing the latest lunacies than did operetta.

The songs in musical comedy hewed close to recitative, unlike those in operetta. As often as not, they could be talked rather than really sung. ("Shouters" was a term many critics employed early on to describe some musical-comedy singers.) Conversely, operetta music was vaultingly lyric until recently, requiring the best singers to do it justice. Musical comedy also remained reasonably steadfast in the nature of its music, just as it kept firmly to its comic approach and changed the essence of its librettos far less than those of operetta.

Imported ballad operas were America's first professional musical entertainments—popular attractions such as John Gay's *The Beggar's Opera* and the now forgotten *Flora; or, Hob in the Well*. Ballad operas were hardly operas. They told simple, often bucolic, stories. Although the songs that commented on these stories had original lyrics, their music was frequently appropriated from already familiar airs. Most were probably sung full-throated, but many of them were probably "talked." Soon, alongside these ballad operas, more elaborate spectacles and genuine English operas found places on American stages. Sometimes opera and spectacle were combined with great commercial if not aesthetic success.

Arguments can be marshalled to suggest that modern-day operetta and musical comedy both had roots in these early ballad operas. More convincingly, the popularity in America of more traditional English operas in the 1840s and 1850s, operas such as *The Bohemian Girl* and *Maritana*, can be seen to have smoothed the way for comic operas of later decades. By the same token, the real seeds of musical comedy were sown even earlier. Pierce Egan's 1823 success, *Tom and Jerry; or, Life in London*, was a precursor of the musical comedies that flooded our stages in later years. This English importation was offered to Americans as "a burletta of fun, frolic, fashion and flash." Its story was simple, little more than a tour of the town. The play was adapted especially for American audiences, although the town being toured remained, somewhat surprisingly, London. After all, the events of 1776 and, more immediately, 1812 must still have been very much alive in playgoers' memories. Evidently, England and its capital still exerted a special cachet in the minds of largely Anglo-Saxon audiences.

The heroes of the piece were Jerry Hawthorn, a country bumpkin, and his city cousin, Corinthian Tom. With Tom's rakish friend, Bob Logic, the boys embark on a few days of painting the town. Their spree takes them to the Burlington Arcade, Tattersal's, Almack's, and other high spots of the London of their day. It also lands them in court after a street brawl. Chastened, Jerry returns home with a village sweetheart, Sue, while Tom's fiancée, Kate, leads him to the altar.

The flimsiest of plots, admittedly, but it was the same basic story that so many of the earliest American musical comedies and revues were to employ. Its simplicity allowed all sorts of opportunities for color, comic situations, and romance. And its loose-jointedness permitted the insertion of song and dance at every turn. The original production's spectacle was modest by later standards, the songs simply put new words to popular melodies, and the waltzes and quadrilles were elementary. Still, healthy seeds were planted and soon were to sprout luxuriantly.

In the 1840s two performers took New York by storm and, in their own ways, paved the road for musical comedy. William Mitchell was an Englishman, Frank Chanfrau an American. Mitchell arrived first. Billed as "the distinguished comedian from Covent Garden," he made his American debut in 1836 but did not hit full stride until he leased the tiny Olympic Theatre at 444 Broadway in December 1839. A well-built, handsome man, with dark, deep-set eyes, Mitchell might well have carved a secure niche for himself in legitimate drama and comedy. Instead, he elected to make a career in farce and burlesque. All through the 1840s his bills were among Broadway's most popular attractions.

Burlesque, of course, was not the burlesque Americans knew a hundred years later. Bumping and grinding chorus girls, strippers, and off-color jokes were unheard of. Nineteenth-century burlesque was smiling buffoonery, far more often than not a lunatic travesty of some beloved play, opera, ballet, novel, or folk tale. At this early date, however, the clichés of late nineteenth-century burlesque had not yet become ritualized. The hero was only rarely played by an attractive young lady, since "trouser roles" were still something of a novelty, and Mitchell frequently portrayed the heroine rather than the role of some comic crone or harridan he might have assumed a few decades later.

In line with the older tradition of ballad opera, the music, as Mitchell readily acknowledged in his playbills, was "begged, borrowed and stolen from all sorts of Opera and Ballets, in the most impudent and free and easy style." Indeed, the fun at Mitchell's Olympic began the moment a playgoer's eye was caught by an Olympic poster, for Mitchell also regularly spoofed the old-style announcements which attempted to attract audiences by capitalizing on salient points. One of Mitchell's read in part:

$500 REWARD

Will be paid to any individual who will satisfactorily prove that the Olympic Theatre will contain ten thousand persons at any one time, or that some of the jokes cracked by Crummles have not been begged, borrowed or

STOLEN
From the veteran Joe Miller, and other venerable punsters
Whose Wit sparkled as brilliantly as
A SPLENDID CLUSTER OF DIAMONDS

Another read:

TERRIFIC ANNOUNCEMENT!

The Public are respectfully informed that
MR. MITCHELL
HAVING ARRIVED FROM EUROPE

Four years ago—has, with very little difficulty, prevailed upon himself to appear on MONDAY next, in a NEW PANTOMIMIC BALLET, entitled

LA MUSQUITOE!
which was never performed at
L'ACADEMIE ROYALE DE MUSIQUE AT PARIS
and it is very probable never will be
PERFORMED THERE 300 SUCCESSIVE NIGHTS!

Having lured patrons in to see this spoof of Fanny Elssler's *La Tarentule*, Mitchell maintained his comic welcome in his programs. Playbills of the era customarily offered staccato plot summaries. So did Mitchell. He gave his audiences some idea of the story they were about to see: "An unexpected occurrence unexpectedly occurs. . . . The plot thickens into the consistence [*sic*] of a London Fog—*Low Retta* [the heroine, played by Mitchell] does all sorts of things—so-do-all the other people." Period playbills also described each character in the cast. Mitchell did too. The Ghost in his parody of *Hamlet* was called "a spirited fellow"; Osrick, "a smart, smart, smart young bachelor"; Ophelia, "her father's daughter"; and Gertrude, "her son's mother."

Unfortunately, no text of any of his offerings seems to have survived. We can, of course, venture an educated guess at what went on, based on the texts of surviving contemporary burlesques and on reviews. Much of his fun was visual. Imagine Mitchell portraying the heroine in a spoof of *La Sonnambula* as "in simpering gladness she engages in soft dalliance with her swain," or picture him later "scrambling over housetops, or descending by pump-handles and water-casks." When the fun was not physical, it frequently—all too frequently for modern tastes—derived from punning. In the poster quoted earlier, Mitchell himself alluded to this ("venerable punsters"). His titles too confirm the high esteem in which wordplay was so long held. His travesty of *La Cracovienne* became *Crack-a-vein*; *Bayadere; or, The Maid of Cashmere* was turned into *Buy-it-Dear, 'Tis Made of Cashmere*.

Every now and then Mitchell produced a burlesque that was not based on established material. One of his greatest successes was *1940!; or, Crummles, in Search of Novelty*. This "local Extravaganza," as Mitchell called it, gave his audiences a tour of their city one hundred years in the future. Clearly, however, Mitchell's author, A. Allan, had written a variation of *Tom and Jerry*.

Another playwright, B[enjamin] A. Baker, used the same frame eight years later for *A Glance at New York in 1848*. First performed on February 15 of that year, the play became, as theatrical historian George C. D. Odell recorded, "one of the greatest successes ever known in the history of the New York stage." With appropriate changes in localities and a minor change of title, it delighted audiences across America. For example, hardly had the show opened in New York when ninety miles away A *Glance at Philadelphia* took to the boards with marked applause. A *Glance at New York* was not an outrageous burlesque, as was almost anything Mitchell had put his hand to, but an amused and amusing excursion through New York high life and low life, with a clear emphasis on the low life. Characters, if larger than life, nevertheless professed to be true to life. Another country bumpkin, George Parsells, comes to the big city to visit a friend, Harry Gordon, and their misadventures, not unlike those of Tom and Jerry, provide the show's substance. The boys were even given romantic interests, although their flirtations were so casual and incidental that the pairings just before the final curtain were probably something of a surprise to the audience.

Unlike Tom and Jerry, however, George and Harry were not the center of attraction. That fell to Mose, "a true specimen of one of the B'hoys," rowdy ruffians of the lower orders who were more or less lawful predecessors of modern street gangs and who nightly packed theatre pits and galleries. Mose was a fireman, complete with "his red shirt, his soaplocks [hair plastered down with soap], his 'plug' hat [top hat], and his boots, into which the trousers were tucked at will." Tough, hot-headed but good-hearted, Mose tried, not always successfully, to get George and Harry out of scrapes unscathed.

Frank Chanfrau's performance as Mose rocketed the young man to stardom. (Legend has it that Baker had modeled his protagonist after a celebrated roughneck, Moses Humphreys, "King of Five Points," and that Humphreys's self-proclaimed reign was overthrown by none other than Chanfrau's brother Henry.) Frank Chanfrau quickly appeared in a number of sequels detailing Mose's further adventures. At one point he appeared in two at once, playing at two different theatres on the same night. Since theatres offered several plays each evening. Chanfrau could play Mose at the curtain raiser in one house and then rush over to the second theatre for a play that came later in the bill. In afteryears, the tall, ruggedly masculine actor remained a top attraction in domestic melodrama, notably *Kit, the Arkansas Traveller.*

A *Glance at New York* was, in many ways, a far cry from burlesque. Gone was the perennial doggerel, laden with puns and polysyllabic nonsense. In its place was more or less traditional dialogue, distinguished from much other stage dialogue of the time by its insistent recourse to contemporary slang. Thus, Mose asks George if he is coming to the firemen's ball.

> *George.* Yes, if Harry will go too. I wonder where he is? I lost sight of him in the fight.
>
> *Mose.* Oh! he's safe enough. There's no real pluck in dem auction chaps. Only come tonight, and I'll show you as gallus a piece of calico as any on de floor.

Other scenes caught the more decorous speech of the day. At one point the men dress up as women to invade a ladies' bowling society. The ladies, by the by, are discovered *"dressed in plain white pants and blue blouses and little black caps . . . all smoking cigars."*

> *Mrs. M.* [Mrs. Morton, the chaperone] Girls, you saw the last three persons who entered?
>
> *Girls.* Yes.
>
> *Mrs. M.* I've a secret to tell you. But promise not to scream, or give the alarm in any way.
>
> *Girls.* We promise.
>
> *Mrs. M.* Well, you must know that these people I suspect to be men.
>
> *Girls.* Men!
>
> *Mrs. M.* And one of them no other than Harry Gordon!
>
> *Mary.* Indeed! how did they get in?
>
> *Mrs. M.* I can't tell for the life of me.
>
> *Jane.* I declare if the stout one isn't cousin George! Let's bother them; they dare not avow their sex.

Songs interlarded the action. The music was still borrowed, but the lyrics were new and appropriate. One scene, at a sleazy bar known as Loafer's Paradise, opened with the bedraggled denizens singing to a tune identified in the text only as an Irish air.

> Here we are, a precious crew, that's always on hand
> For a theft or a frolic, at any man's command;
> And a poor deserted lot, too, of late we have been,
> 'Cos we cannot get a visitor that is at all green.
>> For the green 'uns pay the score
>> That's kept behind the door—
>> When once rubbed off, we can get trust for
>> Just as much more!

If dancing accompanied some songs, the text, which purports to offer "the whole of the Stage Business," gives only one hint of it, when the curtain goes up on the final scene to the whole company "dancing a gallopade." No doubt a few elementary routines nevertheless embellished one or two other numbers. It seems curious today how long it took dancing to find a real place in musical comedy. Although by the 1920s many musicals were perceived foremost as "dancing shows," not until a full century after A Glance at New York did choreography become truly integrated with lyric entertainment.

Even the spectacle dancing that became the rage two decades later could not have been offered, for the show, of course, had no real chorus. Still, in the six members of the "Ladies Bowling Saloon" one may see the unpromising beginning of a chorus line.

Throughout the second and third quarters of the nineteenth century, burlesques and Tom and Jerry offshoots delighted American audiences. They are too numerous and, perhaps, too ephemeral to detail here. For example, James Robinson Planché's delightful little spoofs, pieces such as The Deep, Deep Sea, and King Charming, were played across the land and frequently revived with great success. In New York, the beloved W. E. Burton's troupe gave its audiences theatrical city tours with The World's Fair; or, London in 1851 and Apollo in New York.

One other performer and one other play do merit a more extended look. That performer was John Brougham, a fascinating, all too neglected little giant of his era. Brougham was born in Ireland in 1810. Abandoning his medical studies when financial difficulties arose, he took up acting and made his debut at London's Tottenham Street Theatre in an 1830 production of Tom and Jerry. After several seasons at Covent Garden and the Lyceum, he was brought to America in

1843 to fill the roles once played by the recently deceased Tyrone Power. For the next thirty-five years he was in steady demand as an actor and author. His writings ranged from serious drama to farce to musicals. In fact, by stretching only a little, he could be called the father of the American revue. In 1855 he wrote and starred in a burlesque called *Pocahontas; or, The Gentle Savage*, which became the most popular burlesque of its time. By 1860 *Music and the Drama* could note that "*Po-ca-hon-tas* has become a sort of theatrical deity in New York. It has been invested with eternal life by the public, and whenever and wherever it is produced it is invariably received with plenty of applause and laughter. It will become a classic." Six years later, the leading trade journal of the day, the *Spirit of the Times*, could report that demand for it seemed to be increasing year by year.

Brougham's plot was simple. John Smith and his men are greeted warily when they appear at the Tuscaroran court of "King" Pow-ha-tan. But the king's daughter Pocahontas, bored with life at the Tuscarora finishing school, instantly falls in love with the intruder. Pow-ha-tan is furious, for he has promised his daughter's hand to a "Dutchman," Mynheer Rolff. A card game between Smith and Pow-ha-tan settles matters in Smith's favor.

Like Mitchell, Brougham didn't wait for the house lights to dim to set the mood. His wordplay began in the subtitle—*The Gentle Savage*—and continued as he described his entertainment as "An Original Aboriginal Erratic Operatic Semi-Civilized and Demi-Savage Extravaganza." Because the age delighted in polysyllables, the play began with a "Prolegomena," and because the age relished parody, that prolegomena—"The Song of Pocahontas"—was a takeoff on Longfellow's *Song of Hiawatha*, which had been published earlier in the year.

> Ask you—How about these verses?
> Whence this song of Pocahontas,
> With its flavor of Tobacco,
> And the Stincweed—The Mundungus,
> With its pipe of Old Virginny
> With its echo of the Breakdown,
> With the smack of Bourbon whiskey,
> With the twangle of the Banjo;

The play proper begins at the Tuscarora court with red-skinned courtiers singing the praises of their king, praises in which the king himself joins. The music for the number had come from a popular air called "King of the Cannibal Islands." This use of old musical material continued to be characteristic of burlesque, just as it had been of early ballad opera. *Pocahontas's* programs informed its audiences that the music had been "Dislocated and Re-set" by James G. Maeder, who had culled it from diverse sources: old Irish airs, minstrel songs, grand opera, even "Pop Goes The Weasel."

After the opening number, the punning began in earnest. The very first two lines of doggerel are indicative of all the play's speeches.

> *King:*   Well *roared* indeed, my jolly Tuscaroras,
> Most loyal *Corps*, your King *encores* the *Chorus*.

The italics in the printed text were to make certain readers missed no puns. How completely audiences grasped all the wordplay is moot. In its 1866 review the *Spirit of the Times* complained that for a number of reasons "we fail to catch the countless puns with which the whole play is crammed."

However, though puns prevailed as the fount of most humor, anachronisms and similar absurd dislocations of place also claimed their share of laughs. Smith, for example, ends the first scene with a spoof of the famous French actress, Rachel, who had just enjoyed a successful American tour. Other names dropped in the course of the show included Napoleon, Harriet Beecher Stowe, George Bancroft, *Harper's* magazine, Union Square (then New York's theatrical center), and those of any number of now forgotten politicians.

Whatever the source of the laughter—puns, anachronisms, or merely somewhat barbed comments on everyday life—the humor was essentially sunny. There seems to have been little of the sarcasm, with its undercurrents of jaundice or bitterness, that crept into later librettos. Referring to a recent local scandal Smith remarks:

> See that your city fathers work their best,
> When they're fatigued, why, let them have *arrest*.

Political peccadilloes apparently had not yet become as abrasive as they were to be thereafter. Even racial slurs (although they remained common until the 1930s in libretto after libretto) probably did not offend nineteenth-century audiences, however modern playgoers might react. Nor does a jibe at the growing abolitionist movement seem especially malicious. When Pocahontas spots Smith peeping over her seminary wall, she reacts to this man whose skin is not the same color as hers by exclaiming:

> Who are you?
> Are you a *fugitive* come here to seek
> A railway, underground?

To which Smith replies:

> Not by a sight!
> Alas! I'm only an unhappy *wight*,
> Without a *shade* of *color* to excuse
> *Canadian Agents* here to chalk my shoes,
> Therefore my passage-money won't be figured,
> For on that head Philanthropy is *niggard!*

Another source of comedy, one that remained popular until very recent times, was ethnic—the dialect comedian. Brougham made Rolff a "Dutch," that is, German dialect, comedian, who exclaims:

> Mein cootness gracious, was is das I see!
> Is das mein loafley vrow as is to be?

A developing chorus line was evident in the maidens surrounding the heroine—maidens with such typical burlesque names as Poo-Tee-Pet, Luv-Lie-Kreeta, and Lum-Pa-Shuga. Dancing consisted primarily of jigs, clogs, hornpipes, and marching drills.

Richard Moody, the play's modern editor, has written: "Although Brougham's burlesque may seem trivial and its puns 'corny,' much of it, with a small allowance for changes in taste, is not far removed from the television horseplay of Sid Caesar, Red Skelton, or Jackie Gleason." Indeed, in their relative brevity all the pieces discussed so far would be better suited to an hour slot on television than to the two- or two-and-a-half hour productions modern playhouses offer. Not until after the Civil War did full-length musicals begin to sing and dance across American stages.

The most important was unquestionably *The Black Crook*. The show, which combined an already clichéd tale of a man's selling his soul to the devil with elaborately mounted "fairly" ballets, had little aesthetic merit. Beyond its mélange of melodrama and spectacle, no small part of its attraction to playgoers was its slightly suggestive songs and its line of buxom chorus girls clad in flesh-colored tights. The runaway success of 1866, the musical was the first to play over a year in New York, and it alerted producers and theatre-owners to the potential profits of extravagant lyric entertainments.

A season or two later, imported French opéra bouffe afforded those New Yorkers who understood French a look at full-length works with genuine claims to artistry. At the same time George L. Fox gave American pantomime its palmiest nights with *Humpty-Dumpty*. Despite its English roots, Fox saw to it that his show was filled with American humor and American settings. Pantomime, however, never became as popular in America as it was in Britain. With Fox's death it all but disappeared. Lydia Thompson regaled Manhattan "dudes" with her line of beautiful, hefty English blondes and her English burlesques, and J. K. Emmet embarked on a career in sentimental, song-filled plays, of which *Fritz, Our Cousin German*, was the most durable. Most of Emmet's plays told the same basic story: a man, who is separated in infancy or youth from a relative, recognizes that man or woman when an old song, usually a lullaby, is sung. In fact, Emmet's "Lullaby," written for *Fritz* and reemployed in later shows, is the earliest song from our non-minstrel stage still familiar. One Boston-bred burlesque, the 1874 *Evangeline*, quickly endeared itself to American audiences and remained popular for the rest of the century.

None of these offerings precisely foreshadowed latter-day musical comedy, but most contained hints of what was to come. The British blondes, for example, presaged the chorus line and the implicit or explicit sexuality of musical comedy. And while the pantomimes' stories were timeless and the traditional costuming of their transformation scenes a throwback to late medieval style, most of their settings were contemporary and much of their dialogue topical. Much of the action of *Humpty-Dumpty* took place in front of drops representing famous contemporary New York landmarks—City Hall and the very Olympic Theatre in which its audiences were being regaled—while no little part of the dialogue turned not on child-like matters but on civic corruption of the time and economic problems. There was, however, enough broad slapstick so that children who accompanied their parents to matinees might enjoy the show. This dichotomy of naïveté and sophistication was to persist for generations. Nevertheless, the American musical theatre was coming of age, even if its real maturity lay decades ahead.

2

The Great Burlesques

Coming of age is so often a period filled with false hopes, false starts, and moments of giddy promise that realize little. The American musical theatre's adolescence was no exception. Two late-nineteenth-century burlesques, both perennial favorites in their own era and both now forgotten by all but the most devoted scholars, exemplify one well-paved road down which our lyric stage traveled merrily, only to come up short at a dead end.

Evangeline came first; *Adonis* ten years later, in 1884. Precisely halfway between their premieres, in 1879, *H.M.S. Pinafore* sailed forth to forever change English-speaking musicals. The brilliance of Sullivan's melodies and orchestrations and of Gilbert's lyrics and dialogue instantly made all that had gone before (and much that came after) seem banal and dim-witted, if not downright amateurish. Nevertheless, traditional burlesque retained a loyal following in America almost to the end of the century, and nowhere was that loyalty more apparent than in regular and well-patronized revivals of *Evangeline* and *Adonis*.

The work of two New Englanders, J. Cheever Goodwin and Edward E. Rice, *Evangeline* was a very free theatricalization of Longfellow's popular poem. Goodwin was the first important American librettist. Born in Boston about 1852 and educated at Harvard, he took work as a journalist on the Boston *Traveller*. But his real love was theatre. He met Rice when Rice was printing steamship advertisements. Rice, born in 1849 in Brighton, Massachusetts, had left home while still in his teens to become an itinerant actor. Though he had quickly learned all aspects of his trade, he had abandoned it for steadier work as a printer. However, Goodwin's offer of collaboration proved irresistible. In time Rice gave up writing to become a major producer and is credited with giving first important breaks to many contemporary performers and writers: Henry E. Dixey, Lillian Russell, Fay Templeton, Julian Eltinge, and Jerome Kern.

Evangeline, for all its vogue, was a slapdash affair. Goodwin and Rice played havoc with the poet's original story by transporting the lovers not merely from Nova Scotia to Louisiana but to such faraway, farfetched spots as Arizona and Africa. In contrast to the original, the burlesque's hero and heroine are happily reunited at the end, but not before Evangeline has been pursued by an amorous whale and entertained by a prancing heifer. Goodwin and Rice felt free to add a number of characters. Most famous of all was "The Lone Fisherman," who spent the entire evening peering through his telescope at an often invisible ocean and never once uttering a word. Some of the casting indicated how far American burlesque had moved away from its primitive beginnings with Mitchell and Brougham and toward classic burlesque practice. The hero, Gabriel, was played by a woman—a traditional "trouser role"—while the comic harridan, Catherine, was performed by a man in drag.

Musically, the show was more innovative than inventive. At least it offered a totally original score, not one culled from an established repertoire of popular airs. Rice's music was pleasant, if not especially memorable. It ranged competently over all the expected contemporary lyric forms—waltzes, marches, country dances, dithering little comic melodies. The lyrics were unexceptional. They were replete with the contrived literary metaphors and similes, the grammatical inversions and archaisms commonplace at the time. One, added some seasons after the show first appeared, began:

> My heart feels a new-born emotion,
> It never has known before,
> The turbulent tides of the ocean
> Seem thrilling it to the core,
> Since Gabriel's arms were around me,
> No peace in my mind I've known;
> Some spell in its fetters hath bound me,
> Its magical power I own.

Adding and deleting songs (or, for that matter, scenes) was another of the era's commonplaces. Texts were scarcely sacred.

Yet the changes that came and went never touched the work's basic nature. It trafficked in the lightest-hearted buffoonery, as Mitchell's and Brougham's plays had done. Unlike earlier works, however, *Evangeline* was not meant as an afterpiece or as part of a larger bill. It was a full evening's entertainment. Indeed, thanks to the patience of nineteenth-century audiences, it was, by modern standards, a very full evening, running well over two and a half hours. No doubt the examples of *The Black Crook*, *Humpty-Dumpty*, French opéra bouffes, and any number of other musicals prompted the lengthening. Burlesque, and thus prototypical musical comedy, was manifesting further signs of growth.

H.M.S. Pinafore played its part too. The explosive and pervasive effects of *Pinafore* on our musical stages in general and on comic opera or operetta in particular were evident even to contemporaries. From 1879 on, our musical theatre flourished as never before. The number of musical offerings skyrocketed, although the quality of few, if any, approached that of Gilbert and Sullivan's. Still, a new sense of artistic integrity quickly appeared. Sometimes it was more evident as a forlorn goal than as a hailable reality, but a new standard was indisputably in the air and on the pages of theatrical criticism.

The precise effect of *Pinafore* and its successors on musical comedy as a genre is hard to pinpoint, even in retrospect. It was no more than tangential, suggesting that primitive musical comedies might well learn from operetta's example. On the other hand, operetta had come a long way. American operetta might still be embarrassingly clumsy or naïve, despite the earlier successes of imported opéra bouffe, but it was farther down the road to fulfillment than musical comedy. Musical comedy still did not fully exist even as a conception. By a remarkable coincidence, the same season that witnessed the American premiere of *H.M.S. Pinafore* also welcomed the first farce-comedy and saw the first real flowering of Edward Harrigan. In the long run, as we will see in the next chapter, farce-comedy and Harrigan proved the real progenitors of modern musical comedy. Their importance in no way minimizes the importance of *Pinafore* and its Savoyard cousins, however, nor can they totally deny early burlesque ancestry.

It would be difficult to assert that Gilbert and Sullivan's works had a perceptible influence on *Adonis*, apart from some obvious borrowing, yet *Adonis*'s theatrical assurance and charm suggest that some of the Englishmen's style and integrity rubbed off on its author, William Gill. A hundred years after its premiere *Adonis* remains fun to read and must have been a joy when properly performed on stage. (See Appendix 2 for the complete text of the show.) Indeed, of all the nineteenth-century burlesques, *Adonis* might be revived successfully, at least on an institutional stage, if not on Broadway. The principal deterrent to revival would be the lacunae in the surviving text—the many missing lyrics and the absence of all but one or two indications of what melodies were used. Gill wrote little more than that the music was by "Beethoven—Audran—Suppe—Sir Arthur Sullivan—Planquette—Offenbach—Mozart—Hayden—Dave Braham—[John] Eller—and many more too vastly numerous to individualize, particularize or plagiarize." By using established airs, *Adonis* moved a timid step backward from *Evangeline*'s refreshing willingness to create an original score.

Although Gill did not frame the work as a dream, he called his opus a "Burlesque Nightmare," adding it was a "Disrespectful Perversion of 'Pygmalion and Galatea', 'Marble Heart' and Common Sense." The story was relatively simple and was stretched into a full evening's fun by elementary plot contrivances. The sculptress Talamea, "like most of her sex, is in love with her own creation." That creation is a sensually beautiful statue of Adonis. Though she has sold the statue to a duchess, she is loath to part with it. But the duchess has also fallen in love with the marble and will not accept her payment back. At Talamea's behest the goddess Artea (a mythical mythological figure "invented to suit the fell purpose of the Author") agrees to bring the statue to life. He will then be free to choose between the sculptress and the duchess. Talamea is confident he will favor her. The duchess and her four daughters—Pattie, Mattie, Hattie, and Nattie—are shocked but are more than ever desirous of having the living, breathing beauty for their own. They are opposed in this by the villainous Marquis de Baccarat, who covets the duchess and her wealth for his own nefarious purposes. A villain's purposes are always nefarious, and the marquis assured audiences, "I am a polished villain, and have adorned all the very modern melodrama and society plays in that character—When I smile I always show my teeth, and usually perform all my atrocities enveloped in a dress suit, repressed passion, patent leather shoes, and a buttonhole bouquet. But beneath the surface (*raises vest and discloses pistols and daggers*) Ha! Ha!"

Brought to life, Adonis seems indifferent to the women in the atelier. He spurns Talamea. The duchess hesitates,

> Is it love that tugs me so,
> or tight-lacing, I don't know.

But her daughters have no doubts whatever.

> Lovers all away I cast,
> Mr. Right has come at last.

Adonis will have none of this. Reveling in his new-found freedom of movement, he throws up his hands and exclaims,

> If someone don't hold me down,
> red I soon shall paint the town.

But it is not in the lively center of town that the principals immediately find themselves, rather in the suburbs in front of the cottage of Bunion Turke and his daughter Rosetta. (Although Rosetta announces, "I am the village beauty and weigh one hundred and twenty pounds," she originally was played by hefty Amelia Summerville, and, after Miss Summerville left the cast, by the equally imposing George K. Fortesque in drag.) The marquis appears and is smitten by Rosetta. He decides he will strangle her father and then "fly with her to my chateau in Alaska." But first he must marry the duchess for her money. He proves his determination to be a polished villain by once again showing the audience his concealed pistols and daggers.

The duchess, however, can think only of Adonis. "He is at least a gentleman," she snaps at the marquis. The marquis scoffs: "A block of marble a gentleman—a fellow with no family whatever—Why he hasn't even a father and mother. Perhaps even now some of his connections may be acting as Philadelphia doorsteps." When the duchess continues to rebuff him he is only slightly at a loss: "I'll be revenged! I don't know how just at present, but my time will come later on. The practiced dramatist never fails to give the polished villain a chance— and when my chance comes. . . ." Once again he bares his pistols and daggers.

Rosetta, ordered by her father to steal the neighbor's morning paper, reappears, and the marquis instantly embarks on his courtship. Rosetta too wants no part of him. She would run away, but he grabs her just as her father enters. Appalled at the sight of his daughter "in the grasp of a lordly vilyun," Bunion disowns her. "Go! Starve! Die! I care not!" Rosetta pleads in vain. "No matter," her father replies. "I'm a worthy old miller and it's part of my daily routine to turn my daughter out of doors—No better opportunity than this may occur, and I'm not the man to throw a chance away." But before she goes, Bunion reminds her to be sure to steal the newspaper for him.

The scene changes to a garden where Artea is attempting to console Talamea. She urges the sculptress to accept the fact that Adonis does not love her. "Do the birds and flowers love us?" Talamea retorts. "Yet who shall censure us for ordering quail on toast, or buying 'Jack' roses at 75¢ a bud?" The duchess, her daughters, and their lady friends (the chorus) arrive, all mooning over Adonis. But when he appears—walking so gently "He scarcely bends the tender cabbage leaves on which he steps"—he announces he loves no girl, and all girls. He is such "A Susceptible Statue." (Although the lyric for this song is lacking, the manuscript states it was a "paraphrase of the 'Susceptible Chancellor' from 'Iolanthe.' " It was one of *Adonis*'s show-stoppers.) The marquis comes to challenge Adonis to a duel. Adonis is ready—with two foils, one much shorter than the other. Forcing the shorter foil on the marquis, Adonis wins handily. The marquis stalks off, muttering his time will come and displaying his armory to the audience.

The duchess would continue her pursuit of Adonis, but Rosetta and Adonis catch a glimpse of each other and it is love at first sight. Their conversation is anything but the stammering clichés of most lovers.

> *Ros.*   Oh how beautiful.
>
> *Adon.*   Really—
>
> *Ros.*   Excuse the abruptness of the remark but I couldn't help making it.
>
> *Adon.*   Of course you couldn't—I'm not offended.
>
> *Ros.*   How good of you to say so—you see I am a simple village maiden to whom beauty of your classic style is a staggerer.
>
> *Adon.*   While your bucolic loveliness has stamped its trade mark upon my heart.
>
> *Ros.*   You are as elegantly phraseological as you are physically majestic and we provincials are unaccustomed to such bewildering conjunctions.

Adonis has barely taken hold of Rosetta when her father appears. Appalled at the sight of his daughter "in the arms of a lordly vilyun," Bunion disowns her: "Go! Starve! Die! I care not!" Rosetta pleads in vain. However, Bunion does retreat far enough to ask Adonis if he would support him. Adonis refuses. Bunion then recalls a time gone by: "Twenty years ago upon a dark tempestuous night—when the thunder thundered, the rain rained, and the lightning lightened, seated in my little old log cabin. . . ." But the lovers have no time to listen.

Artea, realizing that Adonis is courting trouble as well as Rosetta, flies in with a magic cane to ward off problems. A problem instantly emerges: the duchess, who has arranged to have Adonis, as her property, locked up in her mansion. Rosetta and all the other girls beg the duchess to relent, but, of course, she is determined. Only a wave of his magic cane allows Adonis to escape with Rosetta.

The second act begins in the country, whence Adonis has fled. Artea, like a wailing Greek chorus, speaks darkly of the future:

> *Art.* The cane I gave him will protect him from bodily harm, but he shall find that in spreading the butter of pleasure too greedily upon the bread of life he is liable to be put off with the oleomargarine of disappointment.
>
> *Tal.* He'll never eat bread and butter when he can eat cake.
>
> *Art.* In the meantime the Duchess will sue him for breach of promise—Rosetta, deceived by the Marquis, will desert Adonis—and the daughters of your enemy disguised as Opera Bouffers will hither come and torture him with the old old melodies we all know so well and they sing so badly.

Adonis enters, diguised as a milkman. (In the original production, Henry E. Dixey, who played Adonis, was dressed as Henry Irving in *Hamlet* and spoofed Irving's interpretation before going into the milkman routine.) Adonis has learned quickly. Pumping water into his cans, he advises the audience: "I serve city customers, and if I gave them the real article I might poison them. Yes it may be bad water, but it never would be bad milk." Rosetta, fearing their pursuers have caught on to Adonis's disguise, suggests he wear one of her dresses and pretend he is a girl. He goes off to change just as the marquis arrives.

The marquis tells Rosetta that Adonis is not the honest man he pretends to be. "He is connected with a gang of determined ticket speculators and he has two wives sitting in the lap of poverty in Paris, Kentucky." Rosetta demands proof, telling the marquis that if he is correct she will marry him instead of Adonis. She will do this because "in all well constructed dramas whenever the villain gives the leading lady proof of her husband's baseness, she always says then I am yours." Exultingly, the marquis grabs Rosetta's hand, just in time for her father to enter and see them. Appalled at his child "in the grasp of a lordly vilyun," Bunion disowns her and starts to tell her of an incident that took place twenty years ago. The marquis chases him away but, seeing the duchess and her daughters approach, decides he must be off, too: "I mustn't stay here, but in the midst of the fete Adonis shall meet his fate in me—I'll serve him with the papers, steal away his bride, denounce him as a dynamiter and then the polished villain will be master of the situation. (*falls over pig*) 'Tis nothing, merely an everyday hog-currence—I'm glad I said that it's so villainous."

The marquis leaves just in time to avoid the duchess, her daughter, and their twenty girlfriends. (In the original, the chorus girls were divided into groups, each group dressed like chorus girls from recent opéra bouffes.) The wedding festivities of Adonis and Rosetta include a group of circus performers. But before the wedding can take place the marquis returns with papers for Adonis Marble "in the name of the commonwealth of Sag-Harbor." Disillusioned, Rosetta calls off the wedding. She marries the marquis.

Walking along a country road Artea tells Talamea what has befallen Adonis: "Guided by my unseen influence he wended his way to a village nearby and sought employment in a store—in one short week so talented a financier has he proved to be, that he has driven his employer into bankruptcy, and been given the post of receiver by the creditors." They are followed by Rosetta and the marquis. Once married, Rosetta has turned out to be mercenary and demanding. But the marquis is not daunted. He will murder both Rosetta and the clergyman who married them, then burn the wedding license. This done, he will marry the duchess. "I shall be," he gloats, "the most polished villain of this and every other age." But Rosetta has overheard him. She is, she warns him, "A simple village maiden, who clad in her innate nobility of soul must ever be a match for the titled imbecility of an effete despotism." The evil marquis has met his virtuous match, indeed.

At his country store Adonis resorts to a number of quick-change disguises to deceive his pursuers and admirers, who are often one and the same. Rosetta, however, recognizes Adonis's voice. She is determined to divorce the marquis and marry Adonis, but Adonis is no longer interested. Bunion arrives, now prepared to disown his daughter for disowning the marquis. He starts to tell of what happened twenty years ago, only Rosetta will not listen.

> *Bun.* Go, you are—(*sees beer on counter*) Ah beer (*goes to drink it and it moves away on slide*) You are no longer a beer of mine. (*Exit*)

The duchess appears, still hoping Adonis will love her. Instead, the marquis comes to continue his courtship. When Adonis enters, the marquis rushes to the telephone to call the police. A flight ensues. Adonis flees, disappearing through a mirror.

Everyone has gathered at Talamea's atelier. When Adonis enters, they all demand he choose. He does. He steps up to his pedestal and tells Artea, "Oh take me away and petrify me—place me on my old familiar pedestal—and hang a placard round my neck:—'HANDS OFF.' " Artea agrees. Adonis climbs upon the pedestal and is returned to stone as the curtain falls.

For its day, *Adonis* had almost everything. Its physical production was praised, so were its pleasant dances. The chorus marches, the comic clogs and simple prancings, were elementary by today's standards, but they were all 1884 audiences expected, except perhaps in a few balletic spectacles. Handsome young Henry E. Dixey became a star, and *Adonis* proved his reliable meal ticket for many years. Amelia Summerville too became a name to reckon with.

Of course, Gill's text can be dismissed as dated, and it is. No one would write this sort of farcical absurdity anymore. But *Adonis*'s libretto is still delightful. The story is consistent and reasonably well knit. If the "modern melodrama and society plays" the marquis alluded to are now hopelessly old-fashioned, *Adonis*'s spoof of them remains valid—the saccharine, ludicrously innocent heroine (whether a city or a village type), the conniving villain, the village bumpkin who knows a long-kept secret were all period stereotypes. How delicious of Gill to persist in hinting that a twenty-year-old secret will tie loose ends together and then never let the cat out of the bag!

But Gill did more than tell an entertaining story well. All the while he was twitting contemporary plays, he was also creating a field day for his great clowns. The day of such marvelous clowns has long passed, but one can imagine what a capital comic, like the last of the breed—Bobby Clark, Bea Lillie, the Marx Brothers—might have done with this essentially unfunny scene:

> *Adon.*   Education you know is the system of finding out what is what.
>
> *All.*   Well what is what?
>
> *Adon.*   Oh well if you don't know what is what—I may as well go on to the next branch of study, which is lightning calculations upon the blackboard—Each one can get a clear insight into my system by watching the different expressions I throw into my left eye. Suppose for instance I were a census-taker asking the age of a young lady, and the young lady were—to say—the Duchess—not to say the Duchess wouldn't give her age *(aside)*—No one would take it—*(aloud)* I should approach the young lady in this way—Madam, what year were you born in *(Duchess whispers)* Quite a young thing! Now by putting down a row of figures on the blackboard in shorthand—adding them up, subtracting these, and dividing by 3, I find the Duchess is exactly 22 years, 4 months and 7 days old.

If Dixey's embellishments on this scene must be forever lost, so too must be the nuances 1884 audiences could bring to topical material. For example, Talamea refers to "Jay Gould on a bust, as yet untouched by chisel." While even twentieth-century playgoers might understand the reference and the wordplay, it must have had a special immediacy in 1884. So would another statue which Talamea unveiled at the beginning of the play, "Dame Columbia washing her dirty boy 'Politics.'" Comparing the statue of Adonis to the new Statue of Liberty must have had a special freshness and pertinence. Then there were the references to modern technological miracles. Seeing Adonis move for the first time, the chorus exclaimed, "Perhaps it's worked, perhaps it's worked by electricitee." And the preposterousness of finding a telephone—first introduced to the public only eight years before at the Centennial Exposition—at an out-of-the-way country store must have held a special lunacy for *Adonis*'s original audiences.

Adonis so delighted New York that it was awarded a singular honor: it was the first musical, the first play for that matter, to run over five hundred consecutive performances on Broadway. Not until a decade later did another musical surpass it. Indeed, a third of a century later only two musicals, *Irene* and *A Trip to Chinatown*, could lay claim to longer runs. *Adonis* toured the country, with occasional returns to New York, until the end of the century.

For all its triumph, *Adonis*, as a burlesque, was merely the best and one of the last of a dying breed. Yet in many ways it also broadcast a notion of the future. Unlike numerous burlesques, which traipsed blithely through distant lands and long-gone times, everything about *Adonis* was brashly contemporary. Its outlook, while essentially cheery and innocent, was often jaundiced and sarcastic as well. Its emphasis was on youthfulness. And its music was designed (or redesigned) largely so that its songs could be "talked." Quite possibly, *Adonis* was responding to the new vogues for farce-comedy and for Edward Harrigan's unique plays, both of which had begun to conquer playgoers' affections five years before, and both of which helped to sound the death knell for burlesque.

3

Farce-Comedy and Edward Harrigan

On August 14, 1875, a man named Nate Salsbury submitted an "original musical comic absurdity" to the Copyright Office in Washington, D.C. His entertainment, which he had offered originally to Chicago three months earlier, was called *Patchwork* and had been written specifically for a small group of performers Salsbury had assembled and christened Salsbury's Troubadours. Salsbury was a gambling man and in later years liked to claim he had entered the theatre with $20,000 won in a poker game. Luckily for Salsbury, $20,000 went a long way in 1875.

Patchwork's theme was hardly original, offering as it did merely a brief glimpse of high life below stairs. Servants gather in the kitchen early one morning while the family sleeps off the effects of the preceding night's masquerade. The servants don the leftover costumes and cavort in an "Amateur Theatrical Performance." Their master's footsteps send them scurrying. At that very time the Vokes Family was regaling audiences with a similar piece, *Belles of the Kitchen*. The similarities did not pass unnoticed by critics, and Salsbury eventually was forced to confess that *Belles of the Kitchen* had provided the inspiration for *Patchwork*.

Some students of musical theatre long have given the Vokes Family a curt bow. Audiences in 1875 may have felt there was a minor but crucial difference between the Vokes Family's entertainments and those of Salsbury. The former may have seemed more sketches with music, the latter musical sketches. Quite possibly a certain chauvinism entered the picture. The Vokes Family was English, Salsbury, American. Whatever the reason, even in the nineteenth century Salsbury was frequently awarded the palm as a forefather of musical comedy, while the Vokes Family was politely waved away. Perhaps when we become more familiar with the stage pieces of this era, the Vokes Family and some of their other competitors also may be awarded a niche in our musical theatre's history.

Patchwork left its first critics and audiences sitting on their hands. Undeterred, Salsbury kept it on the boards for two seasons, patching it here, reworking it there. In time, critics and audiences became more responsive, even if they were never overwhelmed. Growing applause, however, convinced Salsbury he was on the right track.

That track was very narrow gauge. Salsbury's idea for the show was to enlist a band of five multitalented performers to sing, dance, clown, and do whatever else was necessary to sustain a piece that was little more than an extended, elaborate variety sketch. Variety was an early term for vaudeville and just then was rapidly supplanting minstrel shows in public favor. As in later vaudeville, performers appeared in designated sequence, capered through their act, and disappeared. Salsbury proposed a simple but major change in this routine. By creating the thinnest of story lines and assigning each performer a character, Salsbury allowed his actors to display bits and pieces of their talents all through the entertainment as well as to develop the characters somewhat. That development was slight, for Salsbury's earliest stories were little more than excuses for a setting and for the singing, dancing, and clowning that followed. They could have been miniature comic operas, except that they were totally contemporary and domestic, totally giddy, and filled with the lightest order of melody.

The performers that Salsbury selected probably could not have handled operetta's more demanding music—at least not seriously. They were shouters and buffoons of the sort that always filled vaudeville bills and later were to constitute the casts of musical comedy. Many of the clowns had received training in one or more circuses. Among Salsbury's first recruits were his silent partner as producer, John Webster, and Nellie McHenry.

Salsbury apparently recognized that *Patchwork* was not ready for the big time. He never brought the piece to New York. But when he had exhausted all obtainable road bookings, he was ready with a second try. This one he labeled *The Brook; or, A Jolly Day at the Picnic.* His gamble paid off handsomely. After a prolonged tour he offered it to New York on May 12, 1879. It was called a farce-comedy, and the term stuck.

For many years critics were to insist that farce-comedy after farce-comedy—and later musical comedy after musical comedy—had no real plot. They meant no complicated plot, one in which sudden twists and turns lead the story in unexpected directions. Yet until revues discarded story lines entirely forty years hence, some slim narrative thread, if only a slight incident or trifling event, served as a skeleton to be fleshed out with song, dance, and comedy. Just as *Patchwork* had assembled household servants before breakfast, *The Brook* took four guests and their host out on a picnic. In the short first act they meet to make preparations. The longer second act shows the picnic itself, held beside the "I-Flow-On-For-Ever" brook. On the surface, the outing is a disaster. The lid of the coffee container has come loose, and the coffee has mixed with the fish bait; pepper has poured into the ice cream. When, in despair, the picnickers open a trunk supposed to contain watermelons they find it loaded only with theatrical costumes. So they, like the kitchen staff before them, put on an impromptu vaudeville. Their turns include recitations from Shakespeare and imitations of famous actors in famous plays. One guest does an extended comic monologue in which a reporter interviews a Frenchman, an Irishman, and a Yankee on the "Chinese Question." Visual gags abound. A telescope keeps growing longer and longer while an uncorked champagne bottle gushes endlessly. In the very short final act, the picnickers return to the host's home.

Songs were culled largely from popular operas and operettas—*The Bohemian Girl, Giroflé-Girofla, The Princess of Trebizonde, The Grand Duchess, La Fille de Madame Angot*. The excerpts were generally the lightest, less lyrical numbers and were sung not with precision and elegance but with a theatrically careless camaraderie. Only two original songs, the title number and one called either "Oh Charming May" or "Oh Dearest May," were part of the earliest programs.

The characters bore names such as Rose Dimplecheek and Festus Heavysides. Names such as these, which glibly provided a superficial description of the character, had long been commonplaces in comedy and musical entertainments. Yet however artificial and cardboard-like these characters were, they did require the performers to pretend to engage in loose social interaction. This set a tone distinct from the studied formality of operetta or from grotesquely exaggerated minstrel caricatures. Salsbury was setting yet another small precedent for early musical comedy.

The lack of a fully developed story and of three-dimensional figures scarcely ruffled the critics. The *Dramatic Mirror*, reviewing the play's New York premiere, reported that the show was "without plot, character, motif, or, we might almost add, incident, but all the same, very funny." The public agreed, so *The Brook* remained a stable of the Troubadours as long as they survived.

Happy playgoers could patronize return engagements and be certain that much they saw would be new, for Salsbury and his entertainers were constantly refreshing their acts, changing substantial portions of the farce-comedy from season to season. Indifference to textual integrity was one of two other early traits of musical comedy Salsbury helped to establish. Of source, in the case of *The Brook* there was not much text to respect.

Perhaps equally important, no small amount of the songs and jokes employed had not been created with *The Brook* or any other particular early farce-comedy in mind. They were written simply as songs and jokes, or, at best, as songs and jokes for a special actor. Many songs, as mentioned previously, were borrowed from other musicals. When stories grew longer and more complicated, and when whole scores were created for later farce-comedies, interpolations frequently became the order of the night. These interpolations often were brought in at performers' insistence, or, just as often, to bolster pallid material that was the basis of the show.

Inevitably, as farce-comedy's vogue spread, other troupes appeared. Soon William C. Mitchell's Pleasure Party, Willie Edouin's Sparks, and, most popular of all, Edward E. Rice's Surprise Party crisscrossed the country, offering Salsbury's Troubadours competition and playgoers slim but joyful escapism. As always, competition forced these troupes toward bigger, if not necessarily better, productions. Within little more than a year, Edouin's Sparks were presenting *Dreams; or, Fun in a Photographic Gallery*, in which ten performers tell a slight tale framed in a novel flashback. Just a few months later *Revels* employed a still larger cast to play out a relatively involved story that interwove several comic love affairs. Fuller casts, more sophisticated plots—imperceptibly, but with remarkable speed, farce-comedy was evolving into musical comedy. By the end of the eighties, *The Brook* seemed hickishly primitive.

Another man, working in a largely different tradition, sped along that evolution. Edward Harrigan, however, also learned his trade in variety, as had farce-comedy performers, and his own tradition evolved out of it. Harrigan met Tony Hart in Chicago in 1871 and established an act with him. Harrigan wrote clever vaudeville sketches for their turn as well as the dialogue and lyrics for the songs required. Harrigan's father-in-law, David Braham, composed the music. In short order, Harrigan and Hart were the most popular team in contemporary variety.

But Harrigan was not content with short, ever-changing sketches. He envisoned comic plays on a larger scale interspersed with hummable songs, plays that would accurately, yet lovingly and amusingly, depict the life he and his audiences knew best: the boisterous, untrammeled world of immigrants and their young offspring, the world of lower Manhattan tenements. Both of Harrigan's biographers, E. J. Kahn, Jr., and Richard Moody, suggest Harrigan was influenced by Chanfrau's old "Mose" vehicles, which Harrigan may have seen when he was young, and which were remembered fondly for many years.

Harrigan and much of his following were Irish, so he turned to his compatriots for his main themes. But Italians, Jews, blacks, and, most importantly, Germans were also part of his melting pot. He decided to center his efforts on a family called the Mulligans, who already had appeared in a few of his minor sketches. On September 23, 1878, *The Mulligan Guards' Picnic* was unveiled. It was not a full-length show but was far longer than a sketch. Lest the change be too startling to audiences, the bill also contained a standard olio (yet another term for vaudeville). *The Mulligan Guards' Picnic* was enough of a success to prompt a sequel, *The Mulligan Guards' Ball*, which opened on January 13, 1879.

Note that date! Two nights later *H.M.S. Pinafore* premiered in New York, and four months later, *The Brook*. The 1878–79 season marked American musical theatre's most crucial turning point, setting the pattern for all early comic opera or operetta and nurturing the first recognizable saplings of musical comedy. As chance would have it, *Pinafore* was such a polished, masterful work that operetta needed to look no further for inspiration. The piece had risen fully grown out of a sea of trite, haphazard predecessors. The success of subsequent Gilbert and Sullivan gems, other English models, and, a decade or so later, the first truly acceptable American imitations was exemplary. With time, they were bound to influence musical comedy, still so uncertain and so undeveloped. Their well-plotted stories, the studied, if theatrical, characterization of their principal figures, their often careful integration of song and story, and the fact that their songs were written with specific stories and characters in mind all presented would-be musical-comedy writers with commercially and artistically rewarded models that could not be ignored. Those writers simply needed to adjust the style and tone to suit musical comedy, and they would be set.

Yet musical comedy continued to be slapdash and tentative and did not truly find itself until the 1890s. Just what effect Gilbert and Sullivan had on Harrigan, or if he ever saw any of their initial productions, is unclear. Nevertheless, Harrigan quietly advanced the art of musical comedy. A comparison of *The Mulligan Guards' Ball* with the work that is often acclaimed Harrigan's masterpiece, *Cordelia's Aspirations*, will make clear what Harrigan accomplished—and also what he failed to do.

The curtain rose on *The Mulligan Guards' Ball* to disclose a simple dining room with the table neatly set and supper in progress. After describing the scene Harrigan added a single word in the text, "Music." The diners were Dan Mulligan, his wife, Cordelia, and their son, Tommy. Tommy is excited about the upcoming Mulligan Guards' Ball, especially since youngsters his age are beginning to take over the Guards. Dan is understanding, but, remembering that the corps was named in his honor, he cautions Tommy not to be disrespectful of his elders. Dan is especially concerned about two friends, to whom he owes money. "That'll aise me up," he tells Tommy. (Debts were a recurring theme in the show, much as they must have been recurring problems in the lives of Harrigan's audiences.) But Dan is less understanding about another matter. Tommy is in love with Katy Lochmuller, their butcher's daughter. Katy's mother is Irish, but her father is German, and that is too much for Dan. "The Divil a Dutch drop of blood will ever enter this family," he warns his son.

For the moment, however, Dan must temper his anger. The Mulligans have not paid their butcher bill in many weeks, and their debt to Lochmuller has risen to a whopping thirty-five dollars. Dan's restraint is sorely tried by the arrival of none other than Lochmuller himself. The butcher has come to discuss the free meats that are to be his contribution to the ball. He is donating good meats, not the sausages Dan accidently sits on and which Lochmuller assures him "is second hand bolognas, I sweep de meat up mit de sawdust, und sell it to de Italians." Unfortunately, Lochmuller has overheard Dan's disparaging remark about him, so he too is ready for battle. The battle is averted when Lochmuller's son rushes in to tell his father that young hooligans are robbing the shop. The butcher hurries away.

The scene changes to a street where Tommy and Katy plan an elopement during the ball. As they head off, Palestine Puter, a black minister, saunters on stage with Simpson Primrose, a barber. Their black paramilitary corps, the Skidmore Guards, also has scheduled a ball. Plans to hold it at a black hall have had to be canceled after the sheriff seizes the building for non-payment of debts, so Puter has booked an Irish hangout, the Harp and Shamrock—the very place Dan has rented for the Mulligan Guards. The blacks have no sooner gone their own way than Dan and his friend, Walsingham McSweeney, both drunk, stagger on making final plans for the ball and plotting revenge on Lochmuller. Reminiscing about the good old days, the pair launch into "The Mulligan Guard." This song, something of a theme song for the series, had been written and introduced several years earlier. Its popularity remained such that Harrigan could not resist finding an appropriate moment to insert it in this and several other shows. Its chorus ran:

> We shouldered guns,
> > And marched, and marched away,
> From Jackson Street
> > Way up to Avenue A;
> Drums and fifes did sweetly, sweetly play,
> > As we marched, marched, marched in the Mulligan Guards.

The scene changes to Primrose's barbershop, where both Dan and Lochmuller come for haircuts. A contretemps ensues as to who is to be first in the barber's chair. This time the sparring match is averted when one of Katy's friends appears, and Lochmuller goes down on his knees to beg for information about his daughter's romantic interest in Tommy. Precisely at this moment Lochmuller's wife enters the barbershop and misconstrues Lochmuller's action. Dan in his turn falls on his knees to beg Mrs. Lochmuller to stop the budding romance. He is still on his knees when Cordelia arrives and also misconstrues his behavior. Both men are forced to run after their wives to explain. On another street, the Skidmores assemble, singing "The Skidmore's Fancy Ball," in which they proclaim,

> Every coon's as warm as June
> At the Skidmore Fancy Fall.

A crowd gathers at the Harp and Shamrock, where Tommy urges his father to "give de boys a song as a send off." That song was one of Harrigan and Braham's most beloved, "The Babies On Our Block."

Oh, Little Sally Waters
Sitting in the sun,
A crying and weeping for a young man:
Oh, rise, Sally, rise,
Wipe your eye out with your frock;
That's sung by the Babies a living on our Block.

When the Skidmores enter, the booking mix-up is revealed. Harrigan's directions read "*Grand rush to C[enter]*. *Business of fight*. *Women scream*. *Negroes draw razors, general melee when Mr. Garlic [owner of the Harp and Shamrock] enters*." The confusion is resolved by the blacks agreeing to accept an upstairs room. This arrangement quickly led to one of Harrigan's most famous and memorable scenes. Carried away by their enthusiasm, the blacks dance and stomp so wildly that the floor beneath them gives way, and they tumble pell mell onto the Irishmen below. Tommy and Katy take advantage of the moment to run away. Back at the Mulligan home, Dan is persuaded to sing another sentimental song, "The Hallway Door." But his pleasure is cut short when various bill collectors appear. Dan chases them away in time for a final curtain.

Some years later, when Harrigan was at the peak of his fame, a rival producer, A. M. Palmer, dismissed his plays as little more than "a prolongation of sketches." Harrigan disputed this view, seeing his works instead as " 'a continuity of incidents,' with some simple reason for their dovetailing, and each link on the string sustained by some natural motive that calls for the building of the entire stage structure." Certainly, even in his earliest extended works, Harrigan's description was closer to the truth than that of the snobbish, possibly jealous, Palmer. Yet both men talked in terms of plays, never in terms of musicals. Harrigan rarely, if ever, referred to his works as musicals, nor was the term employed on his programs or in his advertising. Nevertheless, musicals they were. And as the plays grew longer, dispensing with olios and eventually becoming full, two and a half hour evenings, the number of songs grew along with them. Moreover, from the very first, the songs Harrigan and Braham offered bore pertinently on the stage business at hand, even when they were carelessly introduced with a "sing us a song" lead-in. That sort of cue had long since become a cliché and was to remain so for several generations.

By the time *Cordelia's Aspirations* premiered on November 5, 1883, Harrigan's art was in full flower. Over the years he not only had developed his abilities but had built up a loving, studied history of the Mulligans and their neighbors. Undoubtedly the most important addition to his roster was the Mulligans' outspoken maid, Mrs. Welcome Allup, or, as she was better known, Rebecca. Tony Hart's delineation in blackface of this part was generally acknowledged as his supreme comedy achievement. But there were many lesser figures whose names suggested Harrigan's ear was attuned to the richness of Elizabethan christenings: the Reverend Jonah Woolgather, Adelaide Foglip, Alexis Canfruit, and Gaspard Pitkins.

Cordelia's Aspirations opens amid the bustle of Castle Garden, shortly after a ship from Ireland has docked. First-class passengers include Cordelia Mulligan. Her brother, Planxty McFudd, and her sisters, Diana, Ellen, and Rosy, are also passengers, although they have traveled steerage class, professing first and cabin classes to have been sold out. They are obsessively jealous of their sister and are determined to claim her wealth. "Cordelia's living in clover," Planxty snarls, "We must be in the field when it's ripe." Admittedly, their lives have been hard. Even their "cow would never rise its head for fear of losing its cud." For her own part, Cordelia is determined to move to greener pastures. She has set her heart on entering society and as a first step has begun putting on airs. She drops badly mangled French at every turn and has even gotten Rebecca to learn some. Simpson Puter, driving a baker's wagon, appears to greet Rebecca, and Dan comes to meet Cordelia. Dan is surprised to see so many "nagurs" at the Garden and warns them, "You have to sleep in the Garden with the rest of the Russian jews." The blacks are not daunted. In fact, they are honored. After all, one of them notes, Castle Garden is "whar Columbus landed." But Dan is more surprised by the changes in Cordelia, who makes no bones about her determination to leave Mulligan's Alley, which suddenly seems "beneath my station."

Bridget and Gustavus Lochmuller come walking down a street,

"richly dressed." Lochmuller is now "the owner of the biggest slaughter house in the city." He and his wife have moved to a mansion in Riverside Park, although Gus sighs, "I would rather live on Avenue A." Bridget's rise further fuels Cordelia's ambitions. She now brings Dan his slippers on a silver tray, prohibits anything as lowly as beer in the apartment, and has purchased a house on Madison Avenue. Dan is baffled and hurt, but cannot bring himself to countermand her.

> No, bekase I love her Mac [McSweeney]. Whatever I feel myself getting angry wid her, my mind goes back to Tipperary where we both carried turf to the same schoolhouse and I fancy I can see her milking the little red cow and myself standing beside her and we talking of the future. Thin the memory of the day we emigrated and the day we landed and the many hard winters I struggled wid her in America, her smiling face was sunshine to my heart. . . .

Dan pours his heart out not only to McSweeney, but to lawyer Ridgeway who happens down the street.

On auction day at the Mulligan's home, "The Mulligan Guard" is heard. "Good old song!" cry the neighbors. The auctioneer tells his assistant to hang the flag out the window, so the auction of the Mulligan's unwanted furnishings can begin. Dan watches bitterly as his dearest keepsakes are put on the block. His anger wells over when the auctioneer reaches one battered object.

> There's a place for the coffee and also the bread,
> The cornbeef and praties, and oft it was said:
> "Go, fill it wid porter, wid beer, or wid ale."
> The drink would taste sweeter from Dad's dinner pail.
> It glisten'd like silver, so sparkling and bright,
> I am fond of the trifle that held his wee bite;
> In summer or winter, in rain, snow or hail
> I've carried that kettle, my Dad's dinner pail.

Dan grabs the pail and storms out to end the first act.

Act Two begins with Cordelia holding a grand reception at her new Madison Avenue home. Only Cordelia seems happy. The suffering of Cordelia's sisters, who find so much as a line of Molière or Shakespeare excruciatingly boring, is small compared to Dan's. He is pained by the stiff collars and tight cravats Cordelia has imposed on him and annoyed at the painting and music lessons he is forced to endure. A gloating, conniving Planxty goads Cordelia to continue pressing Dan. Dan has no recourse but to mortify them. When Planxty noisily upbraids Dan for giving his own portrait of George Washington a mustache, Dan retorts, "I used red, white and blue. His favorite colors." Before Cordelia's horrified guests Dan drinks from a fish tank, grabs the crumb brush to slick down his hair, and climbs on the table to knife a particularly succulent piece of meat.

While Planxty prods Cordelia to torment Dan, he is plotting against Cordelia as well. He plants where Cordelia will discover it a forged letter, which purports to come from a lady with whom Dan has had a brief fling. Cordelia is so distraught that Planxty has no difficulty persuading her to sign all her property over to him. Luckily, lawyer Ridgeway, realizing Planxty's treachery, quietly switches documents so that Cordelia actually signs her property over to Dan. Feeling unloved and humiliated, Cordelia spots a bottle marked "Roach Poison" and drinks the contents. Along with the tumbling down of the Skidmores in *The Mulligan Guards' Ball*, this was probably Harrigan's most famous scene. Dan appears, finds Cordelia unconscious, and cries out:

> *Dan.* Cordelia's dying. She drank the roach poison from the bottle.
>
> *Rebecca.* She drank out of dis bottle?
>
> *Dan.* Cordelia's dead.
>
> *Rebecca.* Drunk.

The "roach poison," it develops, was only Rebecca's secret cache of whiskey.

Act Three takes place the morning after the fete. Planxty, unaware of lawyer Ridgeway's ploy, is lording it over the staff. Cordelia's pretensions have grown ludicrous. When Rebecca chances to mention that she has heard a German band playing on Avenue A, Cordelia asks coyly, "Is there such an Avenue?" Longing for that very street and for the good times he enjoyed there, Dan sends a friend out with an expensive Japanese vase to have it filled with beer. He complains, "I'm to have my breakfast at one o'clock in the day and then a lunch at seven and the great supper in the middle of the night." He pleads with his wife, "Cordelia, I know you saved my money and I know you're trying to elevate me, but I can't forget me neighbors. There's no one up here to sit out on the front stoop and have a glass of beer wid me. There's no barber shops open of a Sunday morning where you could hear the daily news of the week and never fish can I buy from a peddling wagon on a Friday." He also sings of his plight, in a song that Richard Moody (Harrigan's latest biographer) suggests shows that Harrigan had heard Gilbert and Sullivan after all:

> I've her father and her mother
> Wid her sisters and her brother
> A lazy idling loafer a big Corkonian
> Her uncles and her cousins
> And her aunties by the dozens
> All living on the earnings
> Of Daniel Mulligan.

When Cordelia remains unyielding, Dan finally blows his stack. Planxty orders him out of the house, waving the paper Cordelia signed. Lawyer Ridgeway arrives and discloses that he switched documents. The house is Dan's. The revelation brings Cordelia to her senses. Planxty and his sisters are expelled, and Cordelia and Dan prepare to return to Mulligan's Alley. The curtain falls with the cast singing "Wear The Trousers."

There were eleven songs in *Cordelia's Aspirations* as well as three overtures or entr'actes and incidental music. This was nearly three times the number of songs in the original, albeit shorter, *Mulligan Guards' Ball*. Had each of the three acts employed the sort of opening number that already was traditional in operetta and that was soon to become standard for musical comedies, Harrigan and Braham's musical program would have been virtually as large and complete as those of later musical comedies. Perhaps Harrigan's perception of his works as plays rather than musicals prompted his relatively quiet openings. (Even George M. Cohan, who often advertised his attractions as musical plays, plays with music, or comedies with music, often followed this practice a generation later.) Nor were Harrigan and Braham songs always carefully led into, not even in later efforts. Yet the songs themselves continued to have a relevance to the play's dramatic moments. For example, Cordelia's celebrated suicide scene could easily have ended with a blackout after Rebecca's disclosure. Instead, Rebecca and Simpson Promrose seize the bottle and sing "Whiskey, You're The Devil."

Several songs prompted dances, although these dances were rarely compelling or imaginative enough to make critics take notice. As often as not they were light jigs or carefree, seemingly impromptu waltzes that were, apparently, sometimes as casual as they looked. The drills that accompanied the paramilitary groups' marches were also not of an especially high order. For the most part, dancing in these productions was little more than incidental decoration.

The plays—and the songs—were clearly of most importance to Harrigan. Yet his stories were obviously not just the boy-meet-girl sagas that later became musical comedy's standby. They were warmly observed, pervasively funny slices of everyday lower-class life. "Laughter and tears should be the component parts," Harrigan once mused, adding, "The sunshine is not appreciated without the shade." From his first hit to his last (*Reilly and the Four Hundred*), Harrigan held so thoughtfully to this philosophy that his entire body of work has a unity of feeling and style. More than one critic has suggested he was the Dickens of our nineteenth-century musical stage—or at least its Hogarth.

When at the peak of their popularity, Harrigan was in his late thirties and early forties, Hart still in his twenties. Most of their supporting players were also in their thirties. Thus, despite Harrigan's picture of two generations and of social turmoil and ambition, his plays were imbued with a notably youthful ardor. In fact, they regularly displayed a zestiness that must have fatigued some older playgoers. On stage, they were far more raucous than a reading or recounting of them might indicate. Many a night the action was stopped so that Harrigan or Hart might respond to a particularly vociferous gallery. Yet the rough-and-tumble intercourse between the players and the audience was minor compared to the explosive acting on stage.

Sometimes it was contained, as when Tommy in *The Mulligan Guards' Ball* presents his father with an exploding cigar. More regularly, the ruckus encompassed the entire stage. "Melee" and "general melee" were commonplaces in Harrigan's stage directions. Most of his plays contained more than one free-for-all. As such, they were probably another accurate reflection of the life Harrigan's audiences knew well.

But Harrigan depicted far more than incessant social jousting. Lochmuller's remark about the quality of some of his sausages, for instance, confronted a problem that existed until the Pure Food and Drug Act became law decades later. Lochmuller's contempt for the Italians to whom he sold the sausages addressed another difficulty: racial and ethnic antagonisms that still persist but can no longer be shown on stage in quite the humorous way Harrigan handled them. By modern standards Harrigan's figures—even his own Irishmen—frequently touched on caricature. At the same time, there was a truthfulness and compassion to Harrigan's portraits that constantly raised them above caricature and made them a rarity on contemporary stages. The accuracy and credibility of those depictions is a matter for historians to assess. For example, would lower-class whites such as Mulligan and Lochmuller patronize a black barbershop? Was this scene a matter of theatrical convenience or a reflection of accepted practice? Similarly, could Cordelia have afforded a first-class transatlantic passage?

Cordelia's aspirations were the aspirations of many in the audience. Slowly but certainly Irish and Italians and Germans began to leave their tenements in the 1880s and 1890s. The shabbiness and roughhouse antics that characterized tenement life came to have less pertinence for them. Like Cordelia, many were anxious to put even the memory of those earlier days behind them. They aspired to a new elegance and discipline and soon came to want their musical entertainments to reflect their new world. It was not a world that Harrigan himself felt totally at ease in. The disruptive departure of Hart from the stage and the death of Harrigan's eldest son took their toll on the writer and he slowly withdrew from the theatre.

It remained for others to move on with musical comedy. A number of young American talents were contributing their share. But just as *H.M.S. Pinafore* and the Gilbert and Sullivan gems that followed had determined the course of American operetta, at the very moment Harrigan was leaving the theatre the English again sailed over to help redirect and revitalize the American musical theatre. This time, they brought with them not operetta but honest-to-goodness musical comedy.

4

Gaiety

Sweetly skipping,
Truly tripping,
Quaintly quipping,
Here we are,
Pertly prancing,
Ditto dancing,
Gaily glancing,
Tra la la la!
Sweetly singing,
Roses ringing,
Flowers flinging,
Near and far. . . .

Then perhaps I would sing a love ballad,
Which is easy if you have no voice.
To be serious I'd eat Lobster Salad,
Then the song would my lover rejoice.
If the plot is unsuited, no matter,
For of course, there'd be no plot at all.
I should want a dark scene
With the limelight all green,
Ev'ning dress and a light fluffy shawl.

So sang a character named Lady Bickenhall in a 1904 London show, *Sergeant Brue*. Her song was called "Musical Comedy." The lyric named no writers, no stars, no shows, no theatres. It seemed content with generalizations. Audiences most likely were satisfied with these generalizations, recognizing the validity of the impressionistic picture presented.

The lyricist might have been more specific. He might, for example, have mentioned Owen Hall, *Sergeant Brue's* librettist. Hall was well known in the West End by 1904. The lyric might have taken notice of William Edouin (the same Willie Edouin who had led the Sparks), Farren Soutar, Zena Dare, or Sydney Barraclough—all members of the cast and popular entertainers. It might have taken notice of equally popular performers in other musicals. And it might have offered a special bow to a theatre where *Sergeant Brue* was not playing.

That theatre was London's most popular musical house, the Gaiety. There, many people still firmly believe, modern musical comedy was born. In 1904 the Gaiety was at the peak of its success. The theatre itself was only a year old. It had been built to replace an older Gaiety that had stood just a few hundred feet away but had been demolished to make room for a road. John Hollinshead had built and guided that older Gaiety until ill health forced his early retirement in the 1890s. Under his aegis it had been the greatest home of English burlesque. When Hollinshead withdrew from the theatrical arena, burlesque's vogue was waning. At the Gaiety, however, it remained very much alive largely because of two beloved performers—Fred Leslie and Nellie Farren. Then double tragedy struck. Within a few months' time Leslie died suddenly and Farren was permanently incapacitated. Nobody could replace them in the public's affection. The Gaiety's new "guv'nor," George Edwardes, had to find something fresh and different from burlesque if he was to keep the Gaiety's coffers full. Edwardes was not an intellectual; he did not sit down and purposefully plan a new art form. He did, however, recognize that his theatre was being transformed, and he understood basically how to accommodate those transformations. Fine tuning would come by trial and error.

Edwardes realized that audiences in London were changing. Burlesque, with its girls in tights and short skirts and its rough, slapstick humor, was a man's entertainment. Victorian proprieties kept many women away. Yet all around London the theatre was gaining a new respectability. Henry Irving, for example, was luring people into the Lyceum who had forsworn theatrical entertainments. So was Charles Wyndham with his French farces and well-made dramas. Gilbert and Sullivan's good, clean fun had even induced ministers to patronize the Savoy. Husbands and rich roués began escorting their wives and sweethearts. Ladies began to accompany each other to matinees. No one loved the theatre more than Edward, the aging Prince of Wales, and where Edward went, society followed.

That society was the key to George Edwardes's success. Its members disdained the raucous roughhouse of burlesque. They wanted something more dainty, more well-behaved, something perhaps a bit more literate, and, most importantly, something that reflected the world in which they took their tea and carried on their flirtations. By accident or by design, Edwardes found a piece that answered these requirements: it was called *In Town*. He presented the "musical farce" on October 5, 1892, at the Prince of Wales Theatre.

The story, by Adrian Ross and James Leader, was as elementary as those of early American farce-comedy. Captain Coddington, a man-about-town, impetuously invites all the chorus girls of the Ambiguity Theatre to a lavish luncheon. Gay blades of the era were given to such extravagant gestures. Unfortunately, Coddington is a gay but impoverished blade. He has no money to pay for the luncheon. A young lord offers to bail him out on the condition that he be invited too. In what was to become musical-comedy fashion, the young lord's mother and father appear at the party along with other uninvited guests. By the time the dust has settled and the glasses have been cleared, Coddington is engaged to the Ambiguity's leading lady and all the other important characters are happily accounted for as well.

The show's music was undistinguished and hardly memorable. However, the lyrics were almost Gilbertian, a far cry from the drivel in *Sergeant Brue* or in most burlesques. Coddington's theme song ran:

> I'm a terrible swell it is easy to tell
> From my dress and my general deportment:

And I wish to declare that of qualities rare
I've a large and a varied assortment.
I'm at dinners and balls and suppers and halls,
I'm never at home for a minute,
And a "Tableau Vivant" would be sure to go wrong
If they hadn't included me in it:
For I'm the chief and the crown
Of the Johnnies who stroll up and down:
The affable, chaffable, finical, typical
        Man about Town.

In truth, the show was not all that good. But it was something different and, for the time, was reasonably well constructed. Indeed, it was more than well constructed—it was magnificently bedecked. One of Edwardes's masterstrokes was to dress his entire cast in the very latest fashions. Before long the Gaiety shows had become instant style-setters, which gave them an additional cachet and prompted ladies to flock to the theatre to learn what they should be wearing.

The show's dancing was modest. Because Edwardes had discarded tights and short skirts in favor of long, graceful skirts, chorus girls could do little more than sweetly skip, truly trip, and, at most, offer a titillating display of ankle in a gentle swirl.

London's critics were pleased, recognizing that the show was a breath of fresh air. The *Sunday Sun* spoke of *In Town* as an "experiment" and pronounced it a "success with a very big S." The *Sunday Times* characterized it as "a curious medley of song, dance, and nonsense, with occasional didactic glimmers, sentimental intrusions, and the very vaguest attempts at satirizing the modern 'masher.' " The reviewer suggested, however, that because the material was still so tentative the show would have had little chance without its highly skilled interpreters. The *Players*, insisting that despite the evidence that there was no plot at all, nonetheless called the show a "cheery, bright entertainment." The costumes were then described in detail. A fourth critic summed up his reactions: "While *In Town* is a reflex of London life and doings, it also illustrates the spirit that actuates English society all over the country, and embodies the very essence of the times in which we live. The characters are types of the day."

Because so much of *In Town*'s success depended on Arthur Robert's clowning—Gaiety shows found a place for comics to carry on something of the burlesque tradition—and because it was a relatively undeveloped affair, some historians will not hand the piece the honor of being the first Gaiety show. Those historians are not bothered by the fact that *In Town* did not play the Gaiety any more than did two "musical farces" that capitalized on its success, *Morocco Bound* and *Go Bang*, in which Edwardes apparently had an important, behind-the-scenes hand. Oddly enough, the show that most critics do award that honor, *A Gaiety Girl*, also played the Prince of Wales Theatre. But for *A Gaiety Girl*, in 1893, Edwardes first employed the term "musical comedy."

The term was not new. It had been applied indiscriminately for a hundred years or more to a variety of lyric entertainments. But with the success of *A Gaiety Girl* and the shows that followed it, the term caught on as never before and came to stand for a particular type of musical theatre. As a rule—and there were to be numerous exceptions to test but not to discredit that rule—these musicals had contemporary, domestic settings; their dialogue and lyrics clung closely to everyday, colloquial speech; their music was less lyrical, apparently deriving more from recitative than arioso antecedents; and their attitude was more cynical than sentimental and starry-eyed.

Being such early examples, the Gaiety musicals displayed their debts to earlier schools far more than later musical comedies did. The lyric from *In Town* quoted earlier would have been at home in comic opera, as would some of the music from these shows. Late French opéra bouffe made its mark on these shows too. Burlesque comedians were almost always made welcome and found a significant place, sometimes at the expense of a certain logic in events, for while critics were wrong to keep on insisting that many of these musicals had no plots, those plots were often slim and open to adjustment. In this respect, early musical comedy was occasionally indistinguishable from early revue, although revue took longer to catch London's fancy than New York's or Paris's.

London's theatrical tradesheet, the *Era*, insisted "Plot is not the 'strong point' of the libretto of *A Gaiety Girl*. . . . It is sometimes sentimental drama, sometimes comedy, sometimes almost light opera, and sometimes downright 'variety show.' " Granting that Edwardes had

not yet elicited a consistent tone and style from his writers and performers, A *Gaiety Girl* did have a plot, and a stronger one than *In
Town* at that. Haughty Lady Virginia Forrest attends a garden party
given by the Life Guards in the shadow of Windsor Castle. Although
she is there to chaperone some other ladies, she is not above flirting
with several of the titled guests. On the other hand, when several
Gaiety Girls appear, she is the first to snub them. One of these girls,
Alma Somerset, has received a proposal from Captain Goldfield, a
proposal she rejects, fearing her acceptance will hurt his social position and chances of advancement in the Guards. Lady Forrest is appalled at Captain Goldfield's interest in a common actress. To ensure
that he will not ask Alma to reconsider her rejection, Lady Forrest has
her maid slip a diamond comb into Alma's pocket and then accuses
her of stealing. The second act moves to the Riviera. At another party,
this time a masquerade, Lady Forrest is tricked into revealing her mean
ploy. Goldfield and Alma are free to marry.

All in all, A *Gaiety Girl* was far superior to *In Town*. Its music
and lyrics moved still farther away from comic opera, and some of the
names it helped to make famous—such as Owen Hall, its librettist—
thereafter became identified with musical comedy. It also contained
an element that so many later Gaiety shows and other musical comedies would adopt as their stock-in-trade: the workaday heroine who
usually marries into a better or at least a richer world. She was not
quite Cinderella, not a besmudged, unloved slavey, but, as a popular
song was to say a few years later, a respectable working girl. The heroine of *In Town* had been an actress, but little had been made in the
show of class differences. The actress heroine of *A Gaiety Girl*, on the
other hand, confronted and confounded social snobbery. Despite her
difficulties on stage, an actress heroine held a certain glamour in audiences' eyes. Edwardes and his authors understood that, although
many of his patrons were well heeled and possibly even in *Debrett's
Peerage*, much of the audience was more mundanely middle class,
and up in the top balcony clerks and shop girls were paying their hard-
earned pennies for a few hours' pleasant escapism. A sense of immediate identification would come even if the heroine was not a lady or
even an actress but just an idealized embodiment of the young women

who looked down raptly from the top of the house. In Edwardes's next show, his first at the Gaiety to be called a musical comedy, his title and heroine were one and the same, *The Shop Girl*.

This tale of a sweet little shop girl who turns out to be an heiress was a 546-performance hit in London. It was only moderately popular in America, however, for its reputation was tarnished and its success impaired by the wave of anti-British sentiment that swept the country at the time of its premiere. In England, Ivan Caryll's charming, dainty melodies, a harmless libretto, and a sumptuous mounting carried the day. One English critic, not totally dazzled by its lavishness, noted, "So splendidly is the little lady apparelled that occasionally one is tempted to forget she is anything more than a lay-figure, intended for the exhibition of magnificent costumes."

Because *The Shop Girl* was the first Edwardes musical comedy actually to play the Gaiety, those historians who hold reservations about calling *In Town* and *A Gaiety Girl* Gaiety musicals award *The Shop Girl* pride of place. It really is of small account which show is selected. What does matter is that, from 1892 until his retirement in 1914, Edwardes presented London with a steady outpouring of delightful musical comedies, most of which were not only extremely successful but spawned dozens of lesser imitations—such as *Sergeant Brue*.

Sergeant Brue's lyrics caught the essence of these musicals—skipping, tripping, quipping, dancing, singing. They were lighter than air, pink (a favorite color) cotton candy, confectionary fluff. Even that grotesque line, "To be serious I'd eat Lobster Salad," made a point (though it was probably inserted in a desperate lunge for a rhyme). Of course, there was nothing truly serious about these theatrical bonbons, but the mention of so expensive a dish spoke to the cavalier hauteur these musicals affected. It was a hauteur that the bejeweled and tailed audiences in the stalls took for granted and that the less elegantly dressed playgoers in the cheaper seats mistook for genuine class. The Gaiety shows and their better imitators did have a certain class, but it was class derived from conscious theatrical artifice, from an often successful attempt at a consistency in style and tone.

Americans never heard "Musical Comedy" sung in *Sergeant Brue*, for the song was cut before the show reached New York. But reach New York the show did, following in the wake of any number of other English musical comedies, whose arrival helped to put an end to the long reign of comic opera. It could not have been at a more propitious moment.

The demise of early comic opera or operetta came about as much because of the genre's internal exhaustion as for any other reason. Its first great creators apparently had burned themselves out. Several had died, or were soon to pass from the scene. Moreover, with a new century about to dawn, a vigorous, driving America was anxious to embrace anything new. By no small coincidence, the Broadway season that saw the first Gaiety musical comedy arrive on our shores also witnessed the opening of the first successful revue. For the next twenty years these new "up-to-date" genres—musical comedy and revue—dominated American musical theatre.

A *Gaiety Girl* was the first musical comedy importation to appear, opening at Daly's on September 18, 1894. Broadway was delighted but not overwhelmed. The *Dramatic Mirror*'s assessment echoed that of the *Sunday Times* for *In Town*, seeing the work merely as an "indefinable musical and dramatic melange" of "sentimental ballads, comic songs, skirt-dancing, Gaiety Girls, society girls, life guards, burlesque, and a quota of melodrama." Of course, contemporary critics could not put the importance of the Gaiety shows in historical perspective and rightly judged each one on its immediate merits.

Although in the very next season, 1895–96, six more West End musicals crossed the Atlantic, there was no outcry for American writers and producers to elevate their sights and establish a new school of lyric theatre, as there had been after *H.M.S. Pinafore* and later trail-blazing musicals. A few of the English importations were among the biggest hits of their Broadway seasons, but many more were not, and none of their New York runs equaled the long stands they had enjoyed in London. Especially in these years, London hits customarily were awarded longer runs than New York successes. Still, English musicals attracted select audiences wherever they played in this country, and, albeit without much fanfare, did offer exemplars that discerning critics and playgoers could admire and writers could emulate. In later years

Jerome Kern was to insist that Gaiety shows long remained his ideal models of grace, charm, and cohesion.

The extent to which American writers were to emulate these musicals is open to question. The British shows grew out of a strikingly different society and reflected that society's often unique attitudes and tastes. For example, the elegant hauteur so many West End musicals affected was alien to our more democratic notions. London musical comedies were caviar for an elite as far as many Americans were concerned. Besides, Americans had been evolving musical comedies on their own, musical comedies that reflected their own more democratic social evolution. These grew out of farce-comedy, out of Harrigan and his imitators, and out of the vaudeville sketches from which all had sprouted. Naturally, they were Yankee through and through, and as such, they had a much broader appeal to Americans.

No better example exists than A *Trip to Chinatown*. The musical had first been performed in Decatur, Illinois, in September 1890. As part of a prolonged tryout it opened at the Harlem Opera House on December 8, 1890, three weeks before Harrigan offered New York his last success, *Reilly and the Four Hundred*, and nearly two years before the West End premiere of *In Town*. From Harlem it moved back into the hinterlands and did not arrive on Broadway until November 9, 1891. It was an immediate hit. When it closed, it had chalked up a run of 657 performances, a record no musical comedy approached or bettered until *Irene* nearly thirty years later.

The libretto was by Charles H. Hoyt, probably the most famous and successful writer of extended farce-comedies. In his short, brilliant career—he died shortly after he turned forty-one—he wrote sixteen stage hits. Among them were A *Parlor Match*, A *Midnight Bell*, A *Temperance Town*, and A *Milk White Flag*. A *Trip to Chinatown* falls smack in the middle of his career, when he was at the apogee of his powers. No definitive text exists of this or any of his other pieces, for Hoyt was a perfectionist who was forever rewriting his material. When Princeton University published his plays in 1940, it offered two third acts for A *Trip to Chinatown*, one radically different from the other. Nor did Hoyt's texts include the many songs that dotted the show. Their placement remains largely guesswork. Read today, Hoyt's plays come across as wonderful period farces. Still, while no wholly accurate recounting can be made of *A Trip to Chinatown*, a reasonably faithful impression can emerge.

Hoyt's tale—which curiously foreshadows that in *Hello, Dolly!*—begins with Rashleigh Gay and Wilder Daly (a character apparently eliminated from most performances) planning to take their girls to a masquerade ball. Wilder was to escort Rashleigh's sister, Tony. However, the men fear that Rashleigh's guardian, his uncle Ben, will not allow them to go, so they tell him they are taking their dates on a tour of San Francisco's Chinatown. They have invited a Mrs. Guyer along as chaperone. Her letter of acceptance is mistakenly delivered to Ben, who assumes the widow is suggesting a rendezvous at the restaurant she names. Ben sends the youngsters away with his best wishes and sneaks off to the restaurant. Naturally, it is the very one where the men have booked a table. While Ben waits for his assignation, he imbibes liberally. Only after the widow and her wards have left does he learn he missed her. His annoyance is compounded by his realization that in his eagerness he has forgotten his wallet. Later, Ben confronts the youngsters to chide them for their escapade, but they have learned of his in turn.

As usual, more than one critic complained that the musical had no plot. Obviously, it had. It may not have been the most original or involved story—the critics' complaint apparently continued to mean twists of plot rather than basic plot—but it sufficed for an evening's entertainment. What few critics addressed was the nature of the entertainment, for unlike the more abrupt transformation from burlesque to musical comedy that London was soon to witness, American musical comedy had evolved a little more slowly—ever so slowly to the extent that its origins go back to *Tom and Jerry*, Mitchell, and Chanfrau. *A Trip to Chinatown* was, in a sense, *The Brook* grown up.

Could an American critic have paraphrased what an English reviewer was soon to say about *In Town?* Could he have written: "While *A Trip to Chinatown* is a reflex of New York (or San Francisco or American) life and doings, it also illustrates the spirit that actuates American society all over the country, and embodies the very essence of the times in which we live. The characters are types of the day"?

Most likely he would have been comfortable with the first sentence. Hoyt's farce-comedies most certainly illustrated the American spirit, although a critic could probably not safely talk of a single American society in the way the British commentator implied English society was monolithic. Before World War I, European life almost always revolved around its upper classes, and the lower classes aspired to and emulated them. Despite the antics and pretensions of the "Four Hundred," the new, still fluid American upper classes were hardly the be-all and end-all of anything except in their own eyes. In Harrigan's *Reilly and the Four Hundred*, the center of attraction remains Reilly not the Four Hundred.

American farce-comedy and the musical comedy which evolved from it looked to the middle classes for material. Admittedly, they often looked to the most well-to-do among the middle class, but not always. Hoyt's plays dealt with tramps, newsmen, cowboys, politicians—a sweeping panorama of native society, although there are few Irish, Italians, Jews, or blacks in his plots. They were left to Harrigan and others. But the figures portrayed generally shared American characteristics: optimism, ambition, spunk, and an indifference to society's careful, polished niceties. Hoyt's characters, however, frequently bordered dangerously on caricature. They were hardly ever believable flesh-and-blood figures. Their names suggest as much—Rashleigh Gay, Wilder Daly, Welland Strong. Strong, for example, was a one-joke figure. He wandered in and out of scenes complaining about his health, without any real connection to the rest of Hoyt's story for A *Trip to Chinatown*. He was a comic hypochondriac straight out of a vaudeville sketch. Of course, a name like Virginia Forrest in A *Gaiety Girl* indicated that allegiance to older ways also remained in English musicals. By modern musical-comedy standards, the principals of the Gaiety shows were one- or two-dimensional at best. Only Harrigan, from his earliest efforts, had moved beyond such cardboard characterizations.

Musically, Hoyt's shows were several notches below the English shows. As a rule he depended on one mediocre composer, Percy Gaunt, for his melodies, and Gaunt gave him one classic, "The Bowery" in A *Trip to Chinatown*. Gaunt and Hoyt also reworked William

Gooch and Harry Birch's 1878 song, "Reuben And Rachel," into "Reuben And Cynthia," and gave it widespread, enduring popularity. (The chorus begins, "Reuben, Reuben, I've been thinking.") During one road company's Chicago stand, Charles Harris interpolated his "After The Ball." The song became the biggest success of its time and added to *A Trip to Chinatown*'s appeal. In short, out of *A Trip to Chinatown* came three songs that are commonly sung to this day—or three more than all the Gaiety shows left behind. Nevertheless, the Gaiety shows had consistently better music than Hoyt or his contemporaries offered American audiences. Their scores were musicianly and variegated. Their strong points were charm and grace. But the songs simply were not demandingly memorable. The best American songs may have been elementary as music, but they possessed a homey immediacy and such a contagious "sing along" quality that many of them are frequently sung even today.

Over the next two decades or so, a number of thoroughly trained immigrant composers began to write for Broadway, raising the level of its art instantly and appreciably. Gustave Kerker, Ludwig Englander, and Ivan Caryll were among the most productive, although Victor Herbert was unquestionably the finest. Herbert, however, wrote in a grander style. His works, even those he considered musical comedy, would almost all unhesitatingly be termed operetta today.

It fell to Kerker to compose the score for the most popular American musical comedy of the era—the most popular around the world but not at home. *The Belle of New York* opened at the Casino Theatre on September 28, 1897, and New York's belle it most certainly was not. It ran a mere 56 performances—a modest showing even in that era of comparatively short runs. Nor did its post-Broadway tour break any records. But it triumphed in London, racking up 674 performances. Berlin and Vienna applauded it. Paris and Brussels saluted it in 1903 as *La Belle de New York*. Almost everywhere in Europe it has been revived regularly. Paris, for example, welcomed it back in 1953 as *La Belle de mon coeur*. At home it has been virtually forgotten.

A few years before *The Belle of New York* opened, in the spring of 1894, the Casino's management had introduced the revue to Broadway with *The Passing Show*. There previously had been a few tentative, prototypical revues, but they had faded from memory almost as quickly as they had disappeared from the stage. But *The Passing Show* had been called a revue (spelled "review") and was such a hit that it initiated a rage for the genre that has never totally abated. Just as American musical comedy evolved to no small extent from farce-comedy, which in turn derived from American vaudeville, revue came from Paris where it also had evolved from variety or vaudeville.

Unlike variety, in which each performer made a single appearance and presented material specially created for him or her, in revue a cast of performers appeared throughout the show and used material written for all of them by one or two writers. In this respect, revue mirrored farce-comedy. Early revues differed from later ones in that they customarily had some tissue-thin story lines to connect songs and sketches. In this respect too, revue aligned itself to early farce-comedy. Since the more advanced musical comedies at this time also had plots that were regularly perceived as gossamer if not non-existent, it was easy to confuse the genres.

Producers added to that confusion. After *The Passing Show's* success, the Casino's management began to mount at least one revue annually. In 1897 playgoers might have thought the Casino was presenting two. When the "Fourth Annual Review," *A Round of Pleasure*, left after a disappointing summer stand, the Casino announced that for the first time it would offer a revue in the fall. That "Fifth Annual Review" was, of course, *The Belle of New York*. The Casino's motives were transparent. Revues were rapidly becoming the lyric stage's most fashionable attractions. Why not label the newest offering as one? Commercially, the decision made some sense; aesthetically, there were good arguments against it. *The Belle of New York* was hardly the wide-ranging hodgepodge that most revues were. It had a solid, reasonably tight-knit story—a relative rarity even in the period's musical comedies. Many of the characters seemed to be flesh and blood. And virtually all the songs had some palpable connection with the plot. The musical was more or less logically integrated. In several ways it reflected the elevating influence of the stylish Gaiety shows.

Hugh Morton's story centered on Violet Gray, an all too fashion-conscious Salvation Army lass. She sets about mending the extravagant ways of Harry Brown, whose hypocritical reformer father, Ichabod, has disowned him. In gratitude Ichabod attempts to annul his son's marriage to Cora Angelique and force him to wed Violet. Violet, however, knows that Harry and Cora are deeply in love, so she alienates Ichabod by singing a risqué French song. Morton's tale moved up and down Manhattan, allowing the introduction of a number of extraneous characters who captured the color of the town. In more ways than one, *The Belle of New York* was something of a "Gay Nineties" *Guys and Dolls,* even if it was also something of a "Gay Nineties" *Tom and Jerry.*

The show made a celebrity of Edna May, whom producer George Lederer had lifted boldly from the Casino chorus to overnight stardom. At the head of the Casino's fine line of beautiful girls, she made a popular hit of the march "They All Follow Me." "The Anti-Cigarette Society" and "The Purity Brigade," both tongue-in-cheek assessments of contemporary—and enduring—foibles, were almost as well known in their day. The "showy, bizarre ballet," set in Chinatown, was a nightly showstopper.

The double bills offered every fall from 1896 on by Joe Weber and Lew Fields at their bandbox Music Hall also deserve mention. Looking backward and forward at once, they coupled a long burlesque of a current Broadway hit with a short musical comedy. Year in, year out, Weber and Fields played two German-Americans up to their necks in comic misadventures. They were America's consummate dialect comedians and played no small part in helping dialect comedy replace the earlier grotesque comedy.

In stylishness, the mountings of these shows apparently surpassed even those at the celebrated Casino. And Weber and Field's chorus line was considered the most eye-filling in town, forcing the Casino to accept a rather embarrassed second place. With Lillian Russell and Fay Templeton among the regulars, the Music Hall also had Broadway's most glamorous leading ladies. Many a critic would have added that Weber, Fields, Peter Dailey, Sam Bernard, and their associates constituted New York's finest roster of comedians. John Stromberg's music may have been no better than Kerker's, but he did leave behind one still-popular standard from these shows, "Ma Blushin' Rosie."

Dancing in the period's shows was not especially memorable. Despite the success of *The Belle of New York*'s Chinatown ballet, ballet in musical comedy was extremely rare. When it was offered at all, it remained ornamental rather than dramatic. Drills and the most elementary prancing sufficed for choruses. Soloists fell back on jigs, clogs, soft-shoe routines, and, after Loie Fuller, wispy "skirt dances." Choruses, after all, were still selected primarily by looks and not by singing or dancing ability.

It is easy to overestimate the influence of Gaiety shows on contemporary American musicals of the nineties and thereafter, and it is just as easy to overrate the singularly American qualities of our homemade entertainments. American musical comedy of this era was American almost by accident. It was American because it was written mostly by Americans for Americans, because it evolved imperceptibly from older American styles, and because so much of it was set in America and dealt with Americans. There was little flag-waving, very little call for an assertively native musical-comedy style. These musicals were simply entertainment. When they were above average, as was *The Belle of New York*, they were as exportable as Gaiety shows, English comic opera, or French and Viennese operetta. Not until the new century dawned did American musical comedy consciously, and sometimes self-consciously, seek to put a uniquely American stamp on its efforts.

5

American Musical Comedy
Flexes Its Muscles

Same old songs about the same old coons,
Few new words but the same old tunes;
Dashing on before us,
We see the merry chorus,
And everything they do you know's been done;
You hear the same old jokes,
That make the same old hit,
The scenery's just the same
But it's been painted up a bit;
You can ask most any showman,
Inquire of Charley Frohman,
And he'll say there's nothing new beneath the sun.

Nothing new beneath the sun? Young George M. Cohan must have thought musical-comedy audiences were coming to believe just that when he wrote this lyric early in his career. One musical comedy must have seemed as lively, as gaudy, as artistically crude as the next. Yet, in his own way, Cohan hoped to change all that.

In the end, he really did not change things all that much. No small part of the electricity he brought to our musical stage came from his own brazen, singular personality. His broad infectious smile, his "talked" delivery, his exuberant eccentric dancing brought patrons flocking to his shows as much as anything else. In Cohan's musicals, the play was not entirely the thing. In several of his earliest shows his melodic inspiration lent a big helping hand. But with or without Cohan in a starring role, and with or without enduring songs, Cohan's turn-of-the-century musical comedies were consistent in one thing, their proud, jubilant flag-waving. Cohan could not have waved his flag at a better time. If there was little or nothing new on our musical-comedy stages, there was plenty that was new and wonderful all over the country.

America had just come of age. As the nineteenth century gave way to the twentieth, America became a world power. Its factories were sending goods to the four corners of the earth. It became a creditor nation. Cities were rising so high that the word "skyscraper" entered our lexicon. Science, medicine, technology were exploding with discovery. Many of the arts still languished, still embarrassingly derivative and dependent on foreign schools, but the age, exemplified by feisty Teddy Roosevelt, was bullishly certain of its future. The arts would find American answers sooner or later, even so seemingly artless an art as musical comedy assuredly would. And George M. Cohan was certain he could help it along.

George M. Cohan was a child of the American theatre. Though he was neither born in a trunk nor on the Fourth of July, his credentials were highly honorable. His parents, Jerry and Nellie Cohan, were staunch troupers in the day when primitive vaudeville was still called "variety." Their son was born in Providence, Rhode Island, on July 3 (and just maybe July 4), 1878. Soon after, he was carried on stage as a part of his parents' act, branching out a few seasons later as "Master Georgie" in an expanded routine. In time, the Cohans' bit included George's younger sister Josephine and briefly, his first wife, Ethel Levey. Cohan had no formal musical training. He learned his trade as he performed and watched others perform.

Although Cohan remained a top star from early manhood until his death, his most fondly recalled performances and his most enduring writings were largely those from the turn of the century. In the twentieth-century's first decade Cohan was indisputably the right man for his time. Virtually all of his timeless favorites were introduced to Broadway within a four-year period. In 1904 *Little Johnny Jones* included "The Yankee Doodle Boy" (better known as "I'm A Yankee Doodle Dandy") and "Give My Regards To Broadway." Two years later, in 1906, Broadway first heard "You're A Grand Old Flag" in *George Washington, Jr.*; as well as "Mary's A Grand Old Name," "So Long, Mary," and the title song of *Forty-Five Minutes from Broadway*. In 1908 *50 Miles from Boston* appeared and, with it, "Harrigan." Thereafter, Cohan's musical inspiration waned, returning only at intervals and, except for "Over There," never at full strength. In a way, then, his career peaked along with the sort of theatre he best understood, a theatre with mass appeal.

Scholars generally recognize that in these years the American theatrical "road" was at its busiest, although far from its artistic best. By about 1906 or 1907 flickering silent films began to make inroads in small-town and neighborhood playhouses. By March 1908 New York City had four hundred movie houses, no small number of which had recently been marginal live theatres. In May of that year the leading theatrical trade sheet, the *Dramatic Mirror*, carried a large advertisement headlined, "Turn Your Opera House into a Moving Picture Nickelodeon This Summer." Many "summer" nickelodeons remained movie houses when fall arrived. From then on, both the *Dramatic Mirror* and *Variety* regularly published small notices recording that such and such a town—sometimes a surprisingly large one—was suddenly without a legitimate playhouse.

For the most part, the first audiences to be lured away from the theatre were blue-collar patrons, the very "plumber and his lady friend in the last balcony" for whom Cohan had often professed he wrote shows. Cohan was lucky. Hardest hit were the melodramas and inane comedies which early films, however jerky and silent, could reasonably mimic. Without color or sound, however, the period's cinemas were at a loss for song-and-dance entertainments. So Cohan was given a few years' breather, and he took the interval to breathe a special spirit into musical comedy.

Cohan's theatre still drew heavily from the less well-bred, less restrained lower classes and thus remained a theatre that enjoyed (or suffered) a lively rapport between stage and gallery. Perhaps more than any other important figure in the history of the American musical stage, except Edward Harrigan, Cohan wrote with a careful eye to a theatre's upper reaches. By his own admission he performed more to second balconies than to high-priced ($1.50) orchestra seats. Cohan was a good businessman and a hardened realist. Packed, enthusiastic balconies meant little money in the bank compared to well-filled orchestras. Cohan's musicals knowingly compromised the demands of both extremes and included the broader touches, the anti-establishment cuts, and the Irish sentiments (a particular problem for Gaelic artists such as Cohan) looked for by the gallery gods as well as the relative subtlety and stylishness, the more contented approach sought by affluent playgoers. Fortunately, there were attitudes shared by audiences from all parts of the house. Most noticeable perhaps was an all but jingoistic patriotism, intertwined with a pervasive, deep-felt optimism. Deriving from these were other, less obvious, communal perceptions, such as the dream of largely citified audiences for a home in the country, as a permanent way of life or simply as a weekend retreat. Cohan handled these with understanding and élan.

Cohan first made his mark on Broadway after his long apprenticeship in vaudeville by expanding two of his vaudeville sketches as vehicles for his family. *The Governor's Son* in 1901 and *Running for Office* in 1903 both provided handsome meal tickets for his family. They were not, however, very classy shows. Cohan's libretto for *The Governor's Son* was an embarrassingly confused labyrinth. One critic in Boston, where Cohan's fledgling efforts were welcomed far more heartily than in New York, found it "as destitute as a jellyfish of coherent outline." Laments about the feebleness of musical-comedy plots continued to be commonplace for many years, but in this instance the critic was wholly justified. Cohan learned quickly: *Running for Office* had a stronger book.

When Cohan felt the critics were unjustified, he ignored them. A number of traditional reviewers assailed his excessive dependence on slang. Cohan retorted that that was the way his characters would talk could they have come to life off stage. He was not writing "literature," he was creating an entertainment about people with whom his audience could identify. He continued to employ an uncommon amount of slang both in his librettos and in his lyrics. If its often ephemeral vogues have noticeably "dated" his works, they have at the same time imbued them with a unique period charm.

Many critics also railed at what they perceived as an excessive amount of patriotism. Cohan wore his patriotism on his sleeve. Moreover, in a cleverly—and sometimes not too cleverly—disguised way, Cohan musicals were often preachy—chauvinistic parables that nevertheless had the good of his country at heart. His feelings ran deeper than mere flag-waving; he wanted his great country to be even better. He attacked wrongs wherever he saw them—temperance movements, xenophilia, bigotry. To this end, he was happy to gild his philosophic pills.

Excessive slang and an all too patent patriotism were not the only charges critics hurled at Cohan. More than one critic complained that Cohan's musical comedies could better be termed musical melodramas. And to some extent the critics were again correct, but then Cohan himself, as noted before, hardly ever labeled his early shows musical comedies. He was not above pulling out all the theatrical stops: a burglar caught in the act, a will ostentatiously destroyed, a kidnapped heroine, filial defiance. He rarely understated when he could underscore; he clearly had little use for the airy delicacy that the critically acclaimed imported musical comedies affected. In many ways, Cohan was closer to Harrigan's theatre than to the new wave breaking all about him. Yet in the twenty or so years between Harrigan's heyday and Cohan's, musical comedy had begun to take on recognizable traits, and Cohan was quick to adapt those he approved of. Thus, like Harrigan, he sometimes offered only a handful of songs in his shows and, going one step farther, sometimes offered whole acts without musical interruption. And while he still frequently eschewed big opening numbers, dancing became a significant feature in his entertainments. Furthermore, for all Cohan's attention to political and social matters, romantic complications assumed an importance Harrigan had minimized but which Cohan's contemporaries spotlighted.

Cohan's traits were all in evidence in his first real hit, *Little Johnny*

Jones, the saga of an American jockey's unsuccessful attempt to win the English Derby and of his search for his kidnapped sweetheart in San Francisco's Chinatown. He waved his flag proudly in "The Yankee Doodle Boy" and tugged at heartstrings in "Give My Regards To Broadway."

Although somewhat less well-known, *George Washington, Jr.* exemplifies Cohan at his best. Cohan's father played the apparent villain of the piece, Senator Belgrave, an Anglophile who is openly contemptuous of the country he serves. His disdain for America is not his only blind spot—"There are no smart women," he asserts at one point. Belgrave would marry his son George (one guess who played George in the original) to Lord Rothburt's blasé, arrogant daughter, Evelyn. But George has eyes only for Dolly Johnson, the niece of the apple-pie American Senator Hopkins. They are all gathered at Mount Vernon. Hopkins is distressed to hear talk of George marrying Evelyn. He warns Dolly about the gossip, and Dolly, accepting circumstantial evidence, believes it. She becomes very cool to George. Sensing the change in her feelings, George blames his father for it. When Belgrave counters that George is "a disgrace to the name of Belgrave," George replies that his father can take his name back. George will take another, and, in honor of the man whose home he is visiting, he announces that he will hereafter call himself George Washington, Jr. Belgrave berates Hopkins as a troublemaker, but Hopkins has the last word—at least for the first act. "No, you're wrong, Belgrave. The man who used to live there [Mount Vernon] did this."

By most standards, George's impassioned rebellion should have convinced Hopkins that the young man was a suitable match for his niece, but since Cohan needed a second act, Hopkins remains dubious. He advises George that he will not give him his consent to marry Dolly until there is a monument built to George Washington, Jr. Hopkins is not as open and above-board as he seems. The English lord and his daughter are actually frauds, hired by Hopkins to embarrass Belgrave politically. George discovers his secret and, despite their quarrel, alerts his father to Hopkins's chicanery. Shamed, Belgrave acknowledges how narrow-minded he had been. He offers to build a monument to George Washington, Jr., in gratitude. Under the circumstances, Hopkins has no choice but to consent to George's marrying Dolly.

All of Cohan's characteristics were on display from *George Washington, Jr.'s* first curtain. The show begins with a flag-raising—the American flag, of course. The show's opening lines are "Good morning, Mr. Stars and Stripes, good morning! I salute you! I salute you!" They are spoken by a minor character, a black porter. As Cohan quickly has the black man remark, "I know I'm unbleached, but I'm American just the same." Later, when a character addresses the black as "nigger," the black replies, "Look here, Colonel, please don't call me nigger. You know my name." That name was Eaton Ham, a name not merely typical of contemporary musical comedy, but straight out of minstrelsy as well. Cohan's public tolerance may have been ahead of its time, but his writing was steadfastly turn of the century. Much of the material he gave Eaton Ham fell in line with the stereotypical routines accorded stage blacks in other shows. For all his public statements about tolerance, Cohan never wrote with the deep, sincere compassion that Harrigan brought to his dialogue.

Of course, it was not Cohan's purpose to dwell on such issues. Unlike Harrigan, he made his point and quickly moved on. He, for example, characterized what he saw as British snobbery in a single, almost Wildean line. Rothburt informs his daughter that Washington was not an Englishman; he was born in America. "How careless!" Evelyn responds. For a look at American politics Cohan resorted to humor that was perhaps more cracker-barrel than Wildean:

> *Dolly.*   Do the Senators get a big salary?
>
> *George.*   That's all according to what corporation they represent.
>
> *Dolly.*   There are no female politicians, are there?
>
> *George.*   There are lots of old ladies in the Senate.

Politicians and corporations were also on Cohan's mind when he wrote the lyrics. In "If Washington Should Come To Life" he wondered what the first president would think of Morgan and Hearst and political machines. What Washington might have thought of the nation's militaristic fervor did not concern Cohan. Most likely he simply assumed that Washington would share his nation's sentiment. As a result Cohan took for granted that the long gone general would acquiesce in his salute to Norfolk (in Dolly's "I Was Born In Virginia") as the "Home of beauties and war talk."

Mention of political machines and Norfolk reflected Cohan's urban background. He was a city boy through and through (after all, theatres could not exist where there were no cities), and he celebrated the urban growth that engulfed his audiences. This was not the city of posh homes, exclusive clubs, and liveried servants—not really the world of the Gaiety shows—but the everyday city with its "rush and shove," its "motor cars," its "railroads underground," and its "wond'rous big skyscrapers all around." For Cohan, and no doubt for many in his audiences, cities had supplanted "wasted fields and plains and barren lands." He shared his audiences' delight in city things like "The biograph," "The phonograph," and the celebrated comedian Eddie Foy.

But Cohan also realized that his audiences longed to flee the city, for vacation or for good. In "You Can Have Broadway," Cohan spoke of longing for "a little cottage, one I may call my own, in some little country village." But he did so, as many of his followers might have, with only half a heart. In the song's second verse he describes the reactions of a city slicker whose supposed dream comes true. The slicker settles down in "Chester, P. A." (Cohan thus unfairly weighted his argument because the town had an unsavory reputation) and "married a jay, Real female ruben." Before long the slicker is back in New York, ostensibly retracting his words but in truth still dreaming of a rustic utopia. Cohan beautifully caught an ambivalence that still persists. Sweet sentiments were not lacking in his songs. Time and again, however, Cohan punctured them with jabs of cynicism. He ended his song about Washington, "I wonder if he'd try, To never tell a lie."

In Dolly's song, Cohan impudently rhymed "Virginia" with "win yer" (undoubtedly pronounced "win ya"), prefiguring Ira Gershwin, Larry Hart, and Cole Porter at their sauciest. New York critics were not amused—not by the rhyme, not by the impudence, not by Cohan. For years, Cohan remained more welcome on the road than in New York, and more welcome everywhere by audiences than by critics. While loud critical choruses decried his crudities and his excesses, his following at the box office grew steadily. Audiences appreciated what critics ignored or rejected. For all his faults, Cohan brought a curious tonal and stylistic consistency to his shows and in doing so, brought a refreshing American breeze to American musical comedies.

So did Frank R. Adams, Will M. Hough, and Joe Howard. They were offering their wares to Chicago, and no Chicago musical comedies were more successful than those this trio wrote. Before World War I, several large American cities were important theatrical production centers, mounting their own shows and touring them regularly. None was more important than our then second city. Between 1904 and 1910 (roughly Cohan's most productive years), Adams and Hough wrote the book and lyrics and Howard the music for a dozen Chicago musical comedies. All but three were major successes. *The Umpire* (America's first baseball musical) and *A Stubborn Cinderella* chalked up nine-month runs in the Windy City, while the year-long run of *The Time, the Place and the Girl* established a record that was not surpassed for many years. A few traveled as far east as New York, but none was truly successful there.

Let's look at the most successful, *The Time, the Place and the Girl*. The heroes are forced to hide in a sanatorium after an unfortunate, nasty brawl. There one of the men rekindles an old romance with a childhood sweetheart, while his buddy falls in love with a nurse. A quarantine allows them to pursue their courtships at some leisure. Adams and Hough's dialogue leaned heavily on contemporary slang and was saturated with sentimentality. The songs unerringly caught the flavor of the libretto. Two wistful reveries, "Blow The Smoke Away" and "The Waning Honeymoon," were the show's hits, the latter's chorus beginning "Honeymoon, Honeymoon, Wonder why you set so soon."

The team's earliest shows. *His Highness the Bey* and *The Isle of Bong Bong*, resorted to a device *The Sultan of Sulu* had made ragingly popular in 1902, plunking Yankees in an exotic setting and having them come into comic conflict with foreign mores. *The Umpire* exiled its hero to Morocco as punishment for an outrageous call during a game. He manages to redeem himself there at a football game and to win the hand of a lady quarterback. For the most part, however,

Hough-Adams-Howard unfolded their musicals in readily identifiable American settings. Their tone was softer and more sentimental than Cohan's, although critics attacked them too for their slangy dialogue. Unlike Cohan's exulting anthems, their best songs addressed the most gentle emotions. "Blow The Smoke Away" (a pipe smoker's reverie) was one of their bigger hits. "Waning Honeymoon" is still occasionally heard. Most popular of all was "I Wonder Who's Kissing Her Now" from *The Prince of Tonight,* although it was many years before the public learned that Howard had not written the song but had bought the rights to it from another composer, Harold Orlob.

Back in New York, and to a very small extent elsewhere, another group of musical comedies was adding a new note to our lyric stage. These were not the work of a single man such as Cohan or of a tightly knit team such as Hough-Adams-Howard. Their authors were few in number and closely allied by social circumstances, for these were Broadway's first black musicals. Many of the finest black talents of the day—Will Marion Cook, Paul Laurence Dunbar, Cole and Johnson—tried their hands. Since there was no ongoing tradition from which to develop their art, they adapted many accepted white patterns, modifying them to reflect both legitimate black characteristics and, perhaps more important, since these shows were to play for white audiences, white stereotypical perceptions. A typical story dealt with a sharp bunco steerer or a con man who tries to swindle a seemingly slow-witted fellow black only to be given a royal come-uppance by the patsy.

Initially, Broadway's welcome was guarded. Several critics openly expressed fears of racial incidents. In the end, however, most critics were beguiled, not so much by the quality of the librettos and songs but by the infectious dancing and the great comic art of the star of many of these early black shows, Bert Williams. Although Williams died young, exuberant dancing quickly became a hallmark of black shows.

Critics saw Williams and his associates often, for black shows attracted limited audiences and had relatively short runs. Williams's longest run was seven months in London with *In Dahomey.* But the show had run only seven weeks in New York.

Very few of the musical comedies we have discussed had long runs. Except for the hometown success enjoyed by the Chicago shows, these shows were lost in the crush of musical offerings that inundated Broadway and the road. The special interest in them that has developed today has been largely in retrospect. Yet most were not failures. Runs that would seem catastrophically short today were commonplace. Many shows came to Broadway only for limited engagements and then often booked for an entire season on a road many times larger than in recent years. A month or two of good business allowed a show to recoup its investment.

What about the era's other American musical comedies? They were a hodgepodge in every way. There were, of course, favorite themes that popped up repeatedly. Auto and yacht races, frequently with villainous foreigners attempting to thwart American heroes and heroines, provided many an evening of excitement. Witness *The Defender, The Girl at the Helm, The Vanderbilt Cup, The Auto Race,* and *The Motor Girl.* (One amusing bit of trivia about these racing musical comedies is the number of them that had a heroine named Dorothy.) Surprisingly, in an era of rigid proprieties, several shows dealt with marital problems and even divorce. *Marrying Mary* is a choice example. The show is also the result of one of the earliest instances of an important contemporary dramatist moving to the lyric stage, for its author, Edwin Milton Royle, based his libretto on his own play, *My Wife's Husbands.* It treated divorce rather cavalierly, an approach which must have antagonized more strait-laced playgoers at the same time it made the subject palatable to more tolerant patrons. Mary Montgomery has divorced three husbands: a senator, a Mormon, and a lush. All three were stock objects of ridicule in the period's musicals. She is now courted by both a father and his son, the former being no less than the founder and president of the Anti-Divorce League. When all three of her former spouses suddenly appear on the scene, she must choose among the five.

By far the most popular theme was one alluded to earlier: Americans set down in the middle of an exotic culture. The musical that gave the motif its initial impetus, *The Sultan of Sulu,* was a Chicago show that opened there early in 1902 and was embraced by New York the following December. Its titular hero is content with his comfortable, bigamous existence on his little Philippine isle. But that insouciant way of life is endangered by the arrival of the U.S. Navy. Ki-Ram, the sultan, is advised that the United States has bought the island and that changes will have to be instituted in accordance with the U.S. Constitution. Hoping to be able to retain his cozy niche, Ki-Ram courts the stern American lady judge advocate who will be in charge. His sweet nothings begin to melt her icy heart until she realizes he merely proposes to add her to his harem. She would wreak a horrible vengeance, but a naval officer arrives to announce that the Supreme Court has declared "the constitution follows the flag on Mondays, Wednesdays, and Fridays only." Since the navy arrived on an off day, Ki-Ram is free to govern as he pleases.

A story of this sort allowed writers to take potshots at American idiosyncrasies by comparing foreign and American values. In "The Smiling Isle" Ki-Ram described Sulu to baffled sailors in negative terms:

We've not a single college
Where youth may get a knowledge
        Of chorus girls and cigarettes, of poker and the like;
No janitors to sass us
No bell-boys to harass us
        And we've never known the pleasure of a labor-union strike.

In the next several seasons an avalanche of musical comedies presented variations on the basic theme: *The Isle of Spice*; *The Yankee Consul*; *The Forbidden Land*; *The Royal Chef*; *The Sho-Gun*; *His Highness the Bey*; *It Happened in Nordland*; *Fantana*; *The Isle of Bong Bong*; *The Rajah of Bhong*; *The Geezer of Geek*; *The Mayor of Tokio*; *The Tourists*; and, most memorably and enduringly, *The Red Mill*.

Kid Conner and Con Kidder were the Yankee heroes of Henry Blossom's libretto for *The Red Mill*. Stranded without money in a small Dutch town, they quickly find themselves enmeshed in the villagers' romantic difficulties. Gretchen, daughter of the local burgomaster, loves a sea captain, Doris Van Damm. The lovers dwell longingly on wild plans to sail away to "The Isle Of Our Dreams." But Gretchen's stern father wants her to marry the governor and threatens to lock her in the mill if she persists in seeing Doris, even though he is all too familiar with the legend of another young lady who long before had been locked in the same mill only to disappear mysteriously forever. When Gretchen fails to heed the warnings, the burgomaster carries out his threat. From the mill Gretchen sings to the "Moonbeams" of her loyalty and prays for help. Kid and Con contrive to rescue her. The governor arrives, proclaiming "Every Day Is Ladies' Day With Me." Finding Gretchen gone, the governor realizes that he actually loves her aunt, the burgomaster's sister, although the only reason he can give is that he loves her "Because You're You." This alliance with the burgomaster's sister allows Gretchen and Doris to wed. Kid and Con, having wangled passage money, head for a boat that will take them to "The Streets Of New York."

Fred Stone and Dave Montgomery, the era's biggest musical-comedy favorites, enlivened the entertainment with their acrobatic antics, including a fall down a trick ladder and an uproarious prize fight. For the comics, Victor Herbert wrote light, easily singable ditties; for the better-voiced lovers he composed music in more of an operetta idiom.

While turn-of-the-century American musical comedies include both George M. Cohan and Victor Herbert among composers and thus display a sizable range of styles and tones, some generalizations can be made. "Same old songs . . . same old tunes," Cohan complained, meaning that these songs, although original, were written in the same styles that had prevailed for several decades. He was not totally accurate. Ragtime had spread like wildfire since its introduction to Broadway in 1896, bringing with it fresh new sounds and tempos. Hardly a musical comedy opened that did not betray its influence. Even so classical a composer as Herbert included a delightful if neglected ragtime tune, "Go While The Goin' Is Good," in his *Red Mill* score. But generally, Cohan's lament rings true. Most musical-comedy scores were derivative and tepid. How often the era's critics resorted to "reminiscent" to characterize a new show's score! Time after time, stars and producers must have agreed and despaired. With dismaying regularity they injected interpolations to give needed boosts to flabby scores. Popular performers such as Marie Cahill and Blanche Ring won their biggest encores with interpolations. Miss Cahill introduced "Under The Bamboo Tree"; Miss Ring, "In The Good Old Summer Time" and "Bedelia." Miss Cahill's insistence on interpolations cost her a starring role in *It Happened in Nordland* after a row with Herbert. As a rule, however applauded interpolations were, they were scarcely more original or inventive than a show's other songs. They merely had catchier melodies. Most basic scores were often pleasing enough (another favorite adjective of the critics was "tinkly"), but great scores—scores filled with lasting songs—were few and far between.

"Few new words" and "the same old jokes," Cohan continued. Here he was on safer ground. While he and others were seeing to it that many librettos displayed more logic, more cohesion, and a more consistent sense of style and tone than in earlier musical comedies, the improvement in musical books was gradual and erratic.

"Dashing on before us, We see the merry chorus, And everything they do you know's been done." Drills and trippingly light routines continued to be the order of dancing. Tap dancing was introduced in 1903 but was employed only by soloists for the most part. It took several years before choreographers caught onto its electrifying effect when performed by choruses. Ballet, when used at all, was the province of operetta, not musical comedy. The best dancing was often assigned to soloists, for most chorus girls were still picked solely for their beauty. John McCabe, Cohan's biographer, describes how Cohan stumbled on his identifying routine. In doing so, McCabe gives a vivid picture of the era's "eccentric" specialties. An orchestra leader took a number at a far slower tempo than Cohan expected: "Desperately he [Cohan] went into a buck dance, and dragged out the steps in order to accommodate the slower tempo, leaping from one side of the stage to the other instead of remaining center stage as was customary. Laughter stimulated him to exaggerate the steps, and he did a scissors-grinder movement with his arms and legs as he threw his head back in an extravagant gesture of comic strain."

Cohan might well have also called the performers the "same old stars." New or old, great comedians continued to be "grotesques," clowns who relied on preposterous makeup and costuming to accentuate their physical buffoonery. Eddie Foy, Frank Moulan, and the young team of Montgomery and Stone belonged in his category. A few comedians, such as Sam Bernard, Jefferson De Angelis, and Joseph Cawthorn, began emphasizing the sort of dialect comedy that Weber and Fields were making so popular in their unique entertainments. The queens of American musical comedy were the "coon shouters," roistering ladies such as Miss Cahill, Miss Ring, and May Irwin, who belted out their songs to the farthest reaches of a theatre. Good singing was more often reserved for operettas and importations. Cohan himself said, "I don't care if I lose my voice. No one will know the difference."

"The scenery's just the same, But it's been painted up a bit." Shows were not conceived by a single set designer or customer. Two or three set designers, each a specialist in a type of setting (garden, ballroom, etc.), might contribute their talents to a single show. The results were generally a gaudy kaleidoscope of styles.

"Inquire of Charley Frohman, And he'll say there's nothing new beneath the sun." Frohman might have agreed. Except for a minor show called *The Rollicking Girl,* he mounted only imported musicals for Broadway until 1908. England, as we have seen, remained the source of the most chic musical entertainments.

Wonder remains at how suddenly and how prodigally American musical comedy expanded after the turn of the century. Farce-comedy, Harrigan, and Hoyt had done their pioneering only a decade or two earlier. Indeed, from the late nineties until well into the first decade of the new century, producers and audiences alike seemingly had forgotten all about Vienna. They were soon to be reminded how important a theatre town the Austrian capital was.

6

American Musical Comedy
Bucks Viennese Operetta

The Merry Widow turned Broadway's thoughts to Vienna once again, reminding many not-so-old timers of how just a dozen or more years before they had been beguiled by the music of Johann Strauss and Carl Millöcker and Franz von Suppé. It did much more than allow audiences to reminisce. It sent women scurrying for broad-brimmed "Merry Widow" hats; it set Americans waltzing and planted the seeds for the "dancing craze," which in a few years was to have Americans tangoing, fox-trotting, and turkey-trotting as well; and it rekindled the vogue for Viennese operetta, shouldering aside the ascendancy of musical comedy in the process.

The Merry Widow is probably the most successful musical ever written. It almost literally swept the Western world onto its feet, and more than three-quarters of a century after its premiere it is revived regularly and affectionately. Its New York opening night at the magnificent New Amsterdam Theatre on October 21, 1907, was greeted with universal rejoicing. It was praised not merely for its beautiful mounting and fine performances, but as a piece of genuine theatrical art. Its libretto was perceived as taut and sensible, its humor fresh and uncontrived, its characters believable and capable of eliciting sympathy. And then, of course, there was the music, that gorgeous free-flowing well of unforgettable melody. "Coming at the end of an epoch of inane musical comedy—grant that it is at an end!" the *Dramatic Mirror* exclaimed, "—the operetta is twice welcome, on account of its own excellence and because it may start a new era in musical entertainment." Broadway had not seen the last of inane musical comedy, but in some ways *The Merry Widow* did inaugurate a new era in musical entertainment, one that occasionally reached out to touch musical comedy.

For seven seasons, from 1907–8 to 1913–14, Viennese operetta and its best American imitators became the standards by which book musicals were judged. They were seen to have more coherence and more musicianship. Unfortunately, they were often seen to be musical comedy. A blurring of definitions quickly beset the musical stage.

Some of the blame undoubtedly lay with producers. Time after time new shows that were patently operettas were advertised as musical comedies. Producers, who must have understood what they were doing, resorted to this tactic because operetta had been out of fashion for many years, while the Gaiety shows and some better American offerings had given a new cachet to the term "musical comedy." America at the turn of the century was enamoured of anything that was "up-to-date."

Moreover, the new school of operetta differed from older schools. Most noticeable was the change of music, gentler and softer than the florid, frequently pyrotechnic music of nineteenth-century pieces. In addition, there was a fundamental change in settings. Most nineteenth-century operettas had turned to remote lands and remote times to impart an aura of romance. Franz Lehar and his followers wrote scores for librettos set squarely in modern times and in fashionable places. They gave their operettas a determinedly contemporary air that had long been the stock-in-trade of musical comedy.

New-school performers were not the trained, fine singers that operettas had regularly employed. Typically, Donald Brian, the first Danilo, was a song-and-dance man whose dancing and good looks far exceeded his singing ability. For these, and a number of lesser reasons, many a playgoer bought tickets for and frequently sat through a show uncertain whether he or she was watching a musical comedy or an operetta.

Nor can we, looking back, always classify the period's musicals with total assurance. A number of shows straddle the fence between musical comedy and operetta, mixing not only terms cavalierly but styles as well. For instance, one of the best, *Madame Sherry*, skirted the problem of clear classification by calling itself "a French vaudeville." In adopting the term, however, it aligned itself more with musical comedy, since musical comedy was far closer to vaudeville— French or American—than was operetta.

Madame Sherry's setting was contemporary and big city, previously the stamping ground of musical comedy, but new Viennese operettas, as we have said, were increasingly employing similar settings. In Otto Harbach's story, Edward Sherry has deceived his uncle Theophilus, a rich, eccentric bachelor who has devoted himself to classical archaeology. The Venus de Milo's missing arms, Edward insists, "have been the only arms he ever really cared for." Theophilus has been generous to Edward, giving him a large allowance and setting him up as head of The Sherry School of Aesthetic Dancing, a school in the tradition of Isadora Duncan. Unwilling to risk his good fortune, Edward has pretended to accede to his uncle's wish that he marry. At proper intervals he has not only written his uncle about his new wife and their wedding but has announced the birth of imaginary children. To please his uncle still further, he has given his wife and children Greek names.

Of course, Edward is not married, although he has pursued a prolonged attachment to one of his dancing instructors, Lulu. Two visitors appear at the school. One is the son of Venezuelan president Gomez, the other Theophilus's niece, Yvonne. Gomez and Lulu are quickly attracted to one another; so are Yvonne and Edward. Their rounds of casual flirtation are interrupted by the sudden appearance of Theophilus. Edward hastily enlists his housekeeper and two of his pupils to impersonate his wife and children. Theophilus, however, is not taken in. He invites everyone for a trip on his yacht and, when they have pulled away from land, announces they will not return until he learns everyone's true relationships. Lulu and Gomez, and Edward and Yvonne are paired in time for a happy ending.

Deception and mistaken identities were motifs as widespread in operetta as in musical comedy. But the story's treatment in *Madame Sherry*, while frequently romantic, was pervasively light and comic. A gentle cynicism was interwoven throughout. The construction betrayed even more marked allegiance to musical-comedy standards. The plot was not as tightly knit and sustained as operetta plots generally were perceived to be. Indeed, the whole third act was all but superfluous. Theophilus's suspicions surface early and could have been resolved readily by the end of the second act. With no really important new turns of plot, the last act was given over to performers demonstrating their specialties. Elizabeth Murray, who played the housekeeper, was permitted to introduce several interpolations, one of which, "Put Your Arms Around Me, Honey," became as famous as anything in the original score. In short, the action ground to a halt for a series of vaudeville turns, thus possibly suggesting to the producers their description of the show.

The score, by Karl Hoschna, swung back and forth between the lightly mincing tunes of musical comedy or music hall and the more florid, demanding melodies of operetta. "The Birth Of Passion," for example, was a wide-ranging, intensely ardent song that might have been at home in any Viennese offering, while "The Smile She Means For You," whose lyric prefigured a very similar one in "Smiles" a few years later, was essentially Tin Pan Alley. *Madame Sherry*'s most enduring song, "Every Little Movement," was a charming "polka francaise" that could have made itself at home anywhere. It was played throughout the show, each time in a different context and apparently with a slightly varying tempo so that it took on fresh nuances with every reprise. This was a favorite device of operetta all during the 1910–11 season in which *Madame Sherry* appeared. *Naughty Marietta* and *The Pink Lady* both resorted to it effectively. *Madame Sherry*'s appropriation may have branded the entertainment as operetta in many minds, although it could just as easily have suggested that at least a few musical comedies were attempting to be a bit more artful.

Artfulness, however, went flying out the window when interpolations were introduced. Most were voguish Tin Pan Alley rags, the best received of which, "Put Your Arms Around Me, Honey," is still revived.

Madame Sherry exemplifies the sort of musical comedy that ricocheted between genres. It was an exceptional musical bijou, a show that still might be revived successfully if contemporary theatrical criticism could accept an abandoned style of lyric theatre with the same tolerance or even pleasure with which it sometimes welcomes superannuated styles of drama or comedy. Because the show was superior, it was hardly typical. Pinpointing a typical American musical comedy of the time is nigh impossible. In these seven seasons, nearly a hundred American-made musical comedies opened on Broadway or in Chicago. Given the often ambivalent nature of many of the shows and the equally frequent carelessness of definition of genre, no exact count is possible. Confusion existed not merely between operetta and musical comedy, but between musical comedy and revue, and, on a few occasions, between what was American and what was not. Thus a 1910 entry called *Up and Down Broadway* was advertised as a musical comedy, although critics correctly labeled it a revue. Early announcements for a 1909 musical comedy, *The Motor Girl*, suggested it was a reworking of an unspecified French musical. The show that Broadway saw, however, was so totally American that, apart from a Parisian setting, critics could detect not the slightest Gallic tinge.

The variety displayed by these musicals might startle modern playgoers. Many a musical, such as *Up and Down Broadway* or *La Belle Paree*, were revues-cum-musical comedy, usually tying together vignettes and vaudeville specialties with a tour-of-the-city tale. Tom and Jerry were still very much alive, as was the Yankee plunked down in exotic territory, though his vogue was waning. Still, the navy had to come to the rescue in *Funabashi*. Comic strips provided inspiration as witnessed by *Mutt and Jeff*, *The Newlyweds and the Baby*, and *Little Nemo*. The rising suffragette movement could have taken encouragement from the many shows that depicted women triumphing in a man's world, shows such as *Fluffy Ruffles*, *The Yankee Girl*, and *The Wall Street Girl*. Black and black-faced entertainments continued to traffic in the outlandish picaresque, their routines and characters steadfastly stereotyped. The theatre, politics, colleges, and even grammar schools, medicine, fantasy—all supplied grist for librettists' mills. Needless to say, boy met girl, lost girl, and won her back in musical comedy after musical comedy.

Whatever the subject, it was treated in a light and superficial manner. Musical comedy was still interested primarily in music and comedy. The story of a young heiress who gives up her comfortable home to work in the slums sounds like something Broadway or Hollywood or television might present today, complete with ghetto language and left-wing propaganda. *The Charity Girl* told that very story in 1912 terms. Rosemary was the charity girl, and her adventures in the slums centered not on hungry children or rats and roaches or welfare cheating but on her comic encounters with a preposterous clairvoyant. The show's songs were not sardonic or bitter but cute and trite: "Yum Yum Time," "The Magic Kiss," "Those Ragtime Melodies," and "Puppy In Front And Puppy Behind." One song may have pleaded with sincerity, "Oh, come and let us rock you in the cradle of our hearts," but it was scarcely characteristic. Nor did tenement gangs dance with knives and draw blood. Instead, Blossom Seeley and Henry Fink stopped the show with "The Ghetto Glide." Only once did the show attempt to make a genuine social comment, in a song that offended so many that in at least one city a police raid followed and the show was forced to close. The song was sung by a lady of the streets who had contempt for handouts and a unique view of the work ethic. Her philosophy was succinct: "I'd Rather Be A Chippie Than A Charity Bum."

Most of the more successful musical comedies of the time were tailored for the great roster of stars who made Broadway their home. Comedians were especially popular. Joe Weber and Lew Fields (who had gone separate ways after 1903), Montgomery and Stone, Eddie Foy, Sam Bernard, Bert Williams, Joseph Cawthorn—the list was long and distinguished. A majority continued to rely on dialect or ethnic comedy. The most popular, Montgomery and Stone, began as acrobatic grotesques, but recognizing that grotesque clowns were losing their appeal, soon concentrated on acrobatic feats and broad humor.

All the great clowns excelled at brilliant visual touches. In *Piff! Paff!!*
Pouf!!! Foy stopped the show by impersonating a sand castle. Montgomery and Stone brought down the house in *The Red Mill* with their
uproarious prize fight.

Fewer in number but almost as popular were the "red hot mom-
mas," the feisty, strong-voiced ladies best represented by Marie Cahill
and Blanche Ring. Surprisingly, ladies who relied largely on glamour
had less success. None was more in demand than petite, teasing Anna
Held, yet even her career owed much to the promotional genius of
her husband, Florenz Ziegfeld.

Song-and-dance artists were always welcome, George M. Cohan
most of all. Now forgotten performers such as John Hyams and Leila
McIntyre regularly found their names in lights above the title. *The
Merry Widow* had made regimented drills old hat. Irene and Vernon
Castle, and their imitators, introduced suave, intimate ballroom routines. Song-and-dance stars and choruses alike waltzed and tangoed
and fox-trotted. Tap had replaced clog. Some unusual dances enjoyed
their moments in the limelight. Apache dancing, for one, was brought
over from France. In this semi-balletic duet, the male violently mis-
treated his lady. Even *Madame Sherry* included an applauded apache
turn.

With every-increasing visual competition from films, shows became more elaborate. Spectacle was not confined to revues, and choruses grew steadily in number. Choruses of forty or fifty or more became the rule rather than the exception.

"Tinkly" and "reminiscent" remained the most bandied critical ad-
jectives when reviewers came to describe the music. No truly great
composer emerged in these years, although Jerome Kern and Irving
Berlin were interpolating songs left and right and were rewarded with
growing recognition. A. Baldwin Sloane was probably as busy as any
Broadway composer at the time. But no one today remembers any of
his songs, and there is no good reason why anyone should. Yet in
these seven years alone Sloane's musical-comedy credits included *The
Mimic and the Maid, Tillie's Nightmare, The Summer Widowers, The
Prince of Bohemia, The Hen-Pecks, Hokey-Pokey, Hanky-Panky, Roly-
Poly,* and *The Sun Dodgers,* as well as several shows that folded out
of town. Tellingly, Sloane went into semi-retirement in 1913, just as

fine young American composers were beginning to find themselves. A few years later he retired completely.

Since no single show was totally representative of this period's musical comedies, an extended look at the ones produced by Lew Fields between 1907 and 1916 will give a better feeling for the general nature of these entertainments by covering the range of shows at the time. These dates stretch the boundaries of the musicals covered in this chapter a little. However, Fields's 1907 offering, *The Girl Behind the Counter*, which opened three weeks before *The Merry Widow*, was revised and revived in 1916 as *Step This Way*, thereby conveniently framing the series.

In this nine-year period Fields produced twenty-one shows. Besides the two just named, his productions were *The Mimic World, The Midnight Sons, The Rose of Algeria, Old Dutch, The Jolly Bachelors, The Prince of Bohemia, The Yankee Girl, Tillie's Nightmare, The Summer Widowers, The Hen Pecks, The Never Homes, The Wife Hunters, Hokey-Pokey, Hanky-Panky, Roly-Poly, The Sun Dodgers, Marie Dressler's All-Star Gambol, All Aboard, A Glimpse of the Great White Way*, and *Suzi*. Except for *Suzi*, Fields kept hands off Viennese operettas, so his shows exemplified the Broadway musical theatre of his day. *The Rose of Algeria*, a failure despite its magnificent Victor Herbert score, was another operetta, while Herbert's *Old Dutch* and Sloane's *The Prince of Bohemia* were borderline cases. Despite paper-thin plots, *The Mimic World*, the *All-Star Gambol, All Aboard*, and *A Glimpse of the Great White Way* were revues. An argument could be made that *The Girl Behind the Counter* was an English importation, but its libretto was so completely reconstructed that it too becomes a borderline case. It was, however, definitely musical comedy, falling into both the purview of this study and the mainstream of Fields's productions.

In a way, *The Girl Behind the Counter* can be viewed as a star vehicle for Fields himself. Edgar Smith, a sought-after librettist, revamped the English original to spotlight Fields's comic abilities. In his version, Fields, as Henry Schniff, donned a series of ludicrous disguises to thwart his snobbish wife's plan for marrying his stepdaughter to a titled Englishman. Schniff succeeds in tying the knot for his stepdaughter and her American beau. The slack plot allowed Fields to do almost anything, and one thing he wanted was a spectacular scene or two. From the start, his soda fountain scene was a show-stopper. In it, Schniff attempted to match the color of his customers' dress or tie. One particularly difficult young man demanded his striped tie be duplicated. Schniff succeeded, to prolonged applause. Another scene for which Fields held high hopes was a romantic one, a summer arbor at night, bathed in soft lights with additional trick lighting to simulate fireflies. The scene made an excellent first impression, but then fell apart when nothing of interest followed. At the urging of music publisher Edward Marks, Fields interpolated a German song by Paul Lincke, "The Glow-Worm." The song became an immediate hit and played no small part in *The Girl Behind the Counter*'s eight-month run, a very long run for a musical by 1907 standards.

Fields learned two important lessons from the show. The soda fountain scene, and, to a much lesser extent, the arbor scene, confirmed for him the value of spectacle. His future productions were to include some of the most striking and memorable of the era's stage pictures. Fields also learned the value of interpolation. Given the mediocrity of most contemporary Broadway composers, the practice was inevitable, but Fields had countenanced it with some reluctance. After *The Girl Behind the Counter*, he gave interpolation a freer hand.

Three other Fields productions were first and foremost star vehicles—and incontestable American musical comedies to boot. *The Jolly Bachelors*, *The Yankee Girl*, and *Tillie's Nightmare* all opened late in the 1909–10 season. Tiny, abrasive Nora Bayes was the heroine of the first; tiny, assertive Blanche Ring was the Yankee Girl; and the hefty comic harridan, Marie Dressler, played Tillie. Glen MacDonough did the book and lyrics, Raymond Hubbell the music for Bayes's vehicle. MacDonough's story was a little like that of the later *The Charity Girl*. A rich society debutante, bored with her all too comfortable life, takes a job in a drugstore. She prefers the attentions of her lusty middle-class suitors to the thin-blooded playboys of her own class. Unfortunately, her callers' attentions distract her, and she gives one customer the wrong medicine. The rest of the show is spent scurrying through town and a nearby college attempting to retrieve it. Town and gown provided backgrounds for all sorts of theatrical diversions. With Miss

Bayes as the principal lure, the musical ran eleven weeks and then toured.

Fields, remembering the lessons learned from *The Girl Behind the Counter*, included at least one surprising stage effect in *The Jolly Bachelors* and inserted a surefire interpolation to bolster the limp score. The stage effect in this case was an airship sailing across the stage; the song was the still-popular "Has Anybody Here Seen Kelly?" He also employed a small innovation that won high praise but which he failed to follow up on in later shows, for unknown reasons. The drugstore set of the show was done in various shades of brown, and costumes for the scene were carefully coordinated. Other Broadway sets and costumes at the time were generally gaudy, frequently indifferent to tasteless clashes of color.

A pair of established writers, George V. Hobart and Silvio Hein, created *The Yankee Girl* around the special talents of Miss Ring, who was a vaudevillian at heart, most comfortable belting out cheery ditties for two-a-day audiences. She once told an interviewer that plots were merely "something to hang songs on." Hobart obviously accommodated her. Although his story dealt with international intrigue, his treatment was as frothy and inconsequential as *The Charity Girl*'s handling of poverty and social workers was to be. Miss Ring portrayed Jessie Gordon, a smart-as-a-whip young lady who is not about to allow the president of Brilliantina to deprive her father of concessions he has developed in that country. Her nemesis is a Japanese named Oyama (this was the era of the "Yellow Peril"), an industrious, if treacherous entrepreneur who has bribed President Castroaba to rescind the American's grant. Jessie tricks Castroaba into a public admission of his corruption. Her courage in taking on such devious foreigners wins her the hand of a handsome Yankee.

Hobart's dialogue was a pastiche of tried and true jokes, many ethnically slanted, in the fashion of the time. Latin Americans and Japanese were easy targets. "The book," one critic lamented, "shares with the music the dullness of exhausted themes."

Critics were no happier with Hein's melodies. "All the one-finger composers of musical comedies . . . have grounds for action for petit larceny against Silvio Hein," a discouraged reviewer observed. Even

Blanche Ring must have acquiesced. From the very beginning of her stage career she had relied on carefully chosen interpolations to counteract the dreariness of her musical-comedy scores and provide identifiable hits for her vaudeville repertory. She made no exception for *The Yankee Girl*. All through its run she inserted and dropped extraneous numbers, seeking an instantly appealing applause-catcher. She got her biggest hand reprising "I've Got Rings On My Fingers," a song she had introduced in an earlier Fields show. She ranged far and wide in her search, even corralling John Philip Sousa for one number. Hobart and Hein may have tried hard, but their efforts were unimaginative when they were not downright derivative. One of their songs, "Whoop Daddy, Ooden Dooden Day!," was a patent attempt to recapture the flavor and success of "Yip-I-Addy-I-Ay," which Miss Ring had introduced so rousingly in a show two seasons before. "Pretty Polly" reminded at least one critic of another Ring standard, "Bedelia." Best received of the original score was a song called "Top O' The Morning," whose lyric turned Jessie Gordon into an Irish lass who proclaims that "top o' the morning' to you" was as good a greeting as "How do ye do? Glad to see yez, and how be yez?" (Unlike Latin Americans and Japanese, who did not patronize Broadway musical comedies, the Irish went in droves and so were worth saluting.)

The Yankee Girl was a success, running twelve weeks in New York and touring for many months. There was little question that it owed most of its profits to Miss Ring. *Tillie's Nightmare*, not quite as successful, could have acknowledged a similar debt to Miss Dressler. Like Miss Ring, Marie Dressler switched back and forth between the legitimate stage and vaudeville, but she was a clown rather than a singer and had a stronger feel for potential uses of a turn of plot. Nevertheless, *Tillie's Nightmare* had a book that was even looser than *The Yankee Girl's*. Edgar Smith, who had reworked *The Girl Behind the Counter*, brought to life the escapist dreams of a homely, unloved drudge. Years later, in 1926, Fields was to have his son Herbert rewrite the book and set it to songs by Richard Rodgers and Lorenz Hart as *Peggy-Ann*. By that time Freud and psychoanalysis had burst on the scene, and young Fields took clever theatrical advantage of his audience's awareness of them. Matters were different in 1910, how-

Henry E. Dixey as Adonis.
Theatre Collection, Museum of the City of New York

The opening scene of *The Midnight Sons*. The "audience" is actually the chorus (see p. 90).
Theatre Collection, Museum of the City of New York

∽∽∽∽∽∽∽∽∽∽∽

A poster for A *Trip to Chinatown*.
Theatre Collection, Museum of the City of New York

Anna Wheaton and chorus girls sing "Rolled Into One" in *Oh, Boy!*
Theatre Collection, Museum of the City of New York

Patricia Morison and Alfred Drake (center) in the finale of *Kiss Me, Kate*.
Theatre Collection, Museum of the City of New York

∽∽∽∽∽∽∽∽∽∽∽

William Gaxton, Bettina Hall, and Victor Moore in *Anything Goes*.
Theatre Collection, Museum of the City of New York

Sandra Church, Ethel Merman, and Jack Klugman in *Gypsy*.
Theatre Collection, Museum of the City of New York

Stephanie Mills as Dorothy in *The Wiz*.
Theatre Collection, Museum of the City of New York

Jennifer Holliday, Sheryl Lee Ralph, and Loretta Devine in *Dreamgirls*.
Courtesy of Merle Debuskey & Associates

Al Jolson.
Theatre Collection, Museum of the City of New York

Sincerely Yours
Al. Jolson

Marilyn Miller (in *Sally*).
Theatre Collection, Museum of the City of New York

Fred and Adele Astaire (in *Funny Face*).
Theatre Collection, Museum of the City of New York

Ethel Merman (in *Panama Hattie*).
Theatre Collection, Museum of the City of New York

Edward Harrigan.
Theatre Collection, Museum of the City of New York

George M. Cohan.
Theatre Collection, Museum of the City of New York

Lew Fields.
Theatre Collection, Museum of the City of New York

Jerome Kern.
Theatre Collection, Museum of the City of New York

Cole Porter.
Theatre Collection, Museum of the City of New York

Ira and George Gershwin.
Theatre Collection, Museum of the City of New York

Richard Rodgers and Lorenz Hart.
Theatre Collection, Museum of the City of New York

ever. The first International Congress of Psycho-Analysis had been convened only two years before *Tillie's Nightmare* premiered. Modern dream interpretation was the preoccupation of only a small band of theorists and experimenters. The playgoing public as a whole knew little about it, and probably cared still less. Even if they had, Hobart would probably have treated it in a few jokes and moved on. He would not, as Herbert Fields did, have woven it into the fabric of his libretto. Innocent foolery was the show's keynote. Hobart wrote accordingly.

Tillie's dream places her on a luxurious yacht, where she promptly has a hilarious seasick scene. She is invited for an "aeroplane" ride, but her weight impedes take-off. It is all she and the pilot can do to keep her from falling out of the plane. Every incident was an excuse for a bit of broad clowning by Miss Dressler. Curiously, although she was not a singer, it also fell to her to introduce the most famous song A. Baldwin Sloane ever composed, "Heaven Will Protect The Working Girl." Today, the song is recalled only by historians, who might have difficulty naming a second number from the show. The other songs were frequently merely further excuses for the star's comic turns. "What I Would Do On The Stage," for instance, gave Miss Dressler the opportunity to spoof famous dramatic actresses of the day.

At the end of the 1908–9 season Fields produced a musical which he hoped would run through the summer. "Summer musical" was a popular expression of the time, referring to a breezy, brainless entertainment designed solely to take playgoers' minds off the heat. Fields's summer musical, *The Midnight Sons*, succeeded beyond his wildest expectations, running into 1910 and touring to packed houses. Glen MacDonough's book had little to do with that success. It was more typical period frippery, centering ostensibly on the four playboy sons of a senator. The senator, whose monicker, Constant Noyes, was a characteristic bit of musical-comedy nonsense, leaves on a junket to Africa only after he admonishes his sons to find livelihoods or be disinherited. The sons' adventures around the town prompt the jokes, songs, and dances of the show.

Fields gave his production a first-rate cast and another of the gorgeous chorus lines for which he and Weber had long been famous. Vernon Castle, on the brink of fame, was the leading dancer and

nightly won encores for his "amusing, disjointed" stepping. The celebrated director, Ned Wayburn, kept the show moving with his insistence on brisk pacing. However artless and haphazard much of the
entertainment was, it was greeted as a capital "audience show." And
no doubt most playgoers left a performance of *The Midnight Sons*
more carefree than they had been before the show raised its first curtain. Fields began the show with a stunning effect which helped to set
the mood. The real audience found itself gazing at an imaginary audience in an imaginary theatre, "with orchestra, balcony, gallery, and
boxes filled." Indeed, the actual audience found itself on an imaginary
stage, with footlights shining out at them and performers giving them
not so much their all as their backs.

Later in the show, Fields set feet tapping and sent playgoers away
whistling by having Blanche Ring sing the interpolation "I've Got Rings
On My Fingers" over and over again. The song relegated Raymond
Hubbell's basic score to the back seat. Hubbell's music and Mac-
Donough's lyrics were pleasant enough—"tinkly" and "reminiscent."
There were hummable waltzes, Irish ditties, and spoony love songs.
"The Cynical Owl," for example, was a charming rag that told of an
owl, "Once a jovial fowl/Now a cynical birdie," whose home is a tree
at a summer hotel. Every night he hears "Ten or twelve pairs of lov-
ers," but in twenty years "He hasn't heard anything new." The humor
was gentle and smiling, and obviously just right for the time.

The success of *The Midnight Sons* encouraged Fields to mount
several sequels. The first, *The Summer Widowers*, opened on Broadway a month after *Tillie's Nightmare*. MacDonough's story was even
sketchier than the one he had offered for *The Midnight Sons*, recounting little more than the interaction of a band of men (whose wives are
away on vacation) when a glamorous prima donna is thrust into their
midst. (The story was so loose that in some advertisements Fields called
the show a revue.) Sloane's score was pleasant and serviceable. Once
again MacDonough's lyrics were softly humorous if not downright
sentimental and coy. "If you'll furnish yourself and be the missus,"
one lyric concluded, "Gee, I'd like to furnish a flat for you." Mac-
Donough also commented on the ambivalence of contemporary musical comedy, where cynical words and attitudes masked a soft heart.

One character longs to live by the light of the calcium moon that lit musical stages, where "life is naught but the singing of one sweet tune, And romances end the right way."

Fields was unlucky in his search for an outstanding interpolation. Two of his performers—Irene Franklin and Burt Green—at one time or another during the run inserted no fewer than eight songs of their own, all to no avail. Fields even went to Lincke for a second "Glow-Worm," but in the manner of most follow-ups, "Fireflies" was a resounding flop. By way of compensation, Fields provided a sterling cast, which he himself headed, and, of course, remarkable stage effects. One set disclosed an entire apartment house, with its several floors of apartments all bustling with activity. Later in the show, Fields sent miniature airplanes flying through the theatre. A fast-paced, colorful kaleidoscope of entertainment, *The Summer Widowers* was a hit, though it ran only half as long as *The Midnight Sons*. *The Hen Pecks*, by most of the same writers and with many of the same performers, was brought in at mid-season the next year and was equally popular. Fields's spectacular effects for this show included a farm with real animals scurrying about. For the second time running, a superior interpolation eluded him, which may explain the shorter runs of *The Summer Widowers* and *The Hen Pecks*.

The fourth and last in the series, *The Never Homes*, was the least successful, even if in some ways it was by far the best. Certainly it offered the best libretto, a riotous send-up of the burgeoning feminist movement. Militant ladies take over the operation of a city. Their neglected children burn the city down, and the women are remanded into their offspring's custody. The menfolk return to their proper tasks. As always, Fields's production was striking. While gigantic tinsel ribbons, bathed in red and orange lights, simulated the fire, a real fire engine clanked and howled across the stage. This time Fields did discover a delightful interpolation, an Irving Berlin song (though Berlin took no public credit for it), "There's A Girl In Havana." And it was sung in the show by a young Helen Hayes.

But audiences were apparently beginning to tire of these slapdash frivolities. Perhaps Fields was too, although he was loath to admit it. In an interview, while echoing Cohan's earlier lyric, he still insisted

"the public want pretty much the same old thing from year to year."
He did, however, recognize faults endemic in contemporary musical-
comedy construction. "The composer," he observed, unintentionally
demonstrating how randomly songs were integrated into stories, "will
complain that the numbers are not placed properly and the changes
that suit him will probably interfere with the plot. Then the manager
will find fault with the movement of the lyrics because they don't give
him a chance to move the show girls around. . . . And so it goes.
When it is all over consistency of the narrative is a thing that was."

Of course, critics regularly credited the best Viennese importations
with keeping an artful eye on that very consistency. But at the end of
these seven seasons war had erupted in Europe. Vienna was no longer
to supply its distinctive branch of lyric theatre. America would have
to rely more and more on its own talents, whether or not Americans
were willing to create more artful musicals. Happily, a whole galaxy
of native composers and authors were waiting in the wings to do just
that.

7

The Princess Theatre

"It is a pity that you cannot explain or justify delight. Conversation would be so much more amicable if you could. As it is, the friend of your heart or the wife of your bosom, who seems to agree with you on every significant thing in the world, suddenly announces that she cannot abide the Gilbert and Sullivan operettas; and the dead silence that follows indicates that divorce is rapidly setting in."

Thus Gilbert Seldes, the distinguished critic, began a nostalgic trip into a special past—if a look back just nine years could be considered all that nostalgic. Obviously, it was to Seldes. He continued, tongue-in-cheek, "I drag in the Savoy operettas because writing about American musical shows it is always necessary to mention them at least once, and now I have done my duty and can go on with a free heart to the real subject. . . ." Seldes's real subject was a series of shows presented during World War I at a tiny playhouse on the very edge of Broadway's theatre district. That theatre was the Princess, and the musicals that took audiences' minds off the carnage across the Atlantic have ever since been known as the Princess Theatre shows. They were very special musicals indeed, for not only were they the best musical comedies of their day, but they brought American musical comedy into the twentieth century.

Seldes was a careful, sensitive writer, and it was no mere chance that he opened his essay for *Theatre Magazine* by commenting on the delicate nature of delight. There have never been more delightful musical comedies than the Princess Theatre shows. And, while Gilbert and Sullivan's English comic operas may have been a far cry from the American musical comedies Seldes was about to analyze, the two types of shows shared some unique, admirable attributes. Both were incredibly melodic. Their music remains popular. Both were literate and witty. In fact, the Princess Theatre musicals offered the first truly literate, witty body of lyrics the American stage had yet heard. Perhaps less immediately obvious was their shared consistency of style and tone, although their style and tone were disparate. Like the Savoy operettas, the Princess Theatre musical comedies were brilliant, carefully polished gems.

Seldes's affectionate essay was prompted by a 1924 attempt to revive the Princess series. The show, *Sitting Pretty*, did not even play the Princess Theatre and, in any case, was an unfortunate failure. Yet for a few brief moments it filled aficionados with glowing expectations, carrying them back happily to another decade.

For the Princess Theatre that decade began in 1912 when the Shuberts, William A. Brady, and Arch Selwyn announced they would construct a bandbox theatre on the south side of 39th Street, even then near the southern limits of the theatre district. It was to be named after an earlier Shubert house and to be designed by William A. Swasey, who had designed the Shubert flagship, the Winter Garden. From the street, Swasey's playhouse was not particularly inviting, a simple six-story facade that, but for its electric sign, might have passed for a small apartment house. Four urn-flanked doors led playgoers into a low-ceilinged, tile-floored lobby with homey lamps and a fake fireplace. The auditorium was another matter. It was done in blue and off-white in a vague Louis XV style. The house held a mere 299 people, since a number of stringent laws applied to any theatre seating three hundred or more. Almost all the seats were in the orchestra, where an aisle ran down the middle of the house in continental fashion. The balcony had only two rows, meaning that few cheap seats would be available and that the playhouse would be somewhat elitist by default. Because no one apparently envisioned the theatre serving musicals, the stage was the smallest of any professional theatre in New York and the backstage facilities were meager. This ultimately had an effect on the musicals that did play there.

Plans for the theatre were vague at first. Announcements suggested that it was to be used for children's classics or for Grand Guignol. Just before the theatre opened in March 1913, it was decided to present experimental one-act plays, especially those of untested writers. When bills of short plays failed to attract, a young producer named Ray Comstock, associated with the theatre's management, called in a celebrated agent, Elisabeth Marbury. Together they decided to gamble on an intimate musical.

Theirs was the right gamble at the right moment. Lacking the wisdom of hindsight, Comstock and Marbury had no way of knowing that World War I was to have a deadening effect on the creativity of Europe's musical stages, nor that the disenchantment with things European would explode with the war. Since the war had begun—it was the 1914–15 season—and sentiment against the Central Powers had begun to appear, the vogue for Viennese operetta was at an end. Yet it apparently never occurred to either Comstock or Miss Marbury to try a totally American musical. Broadway was still under the sway of foreign influences, and, with sympathy for Britain growing, it seemed only logical to turn to the congenial West End stage. The musical they chose to adapt was Mr. Popple of Ippleton. The men they chose to make the adaptation were Joseph Herbert and Jerome Kern. Herbert withdrew or perhaps was dismissed almost at once, and at Kern's suggestion Guy Bolton was signed as his replacement.

Jerome Kern was born in New York on January 27, 1885, the son of an immigrant father and an American mother. From childhood, music was his sole passion, and he began composing for the public while still in high school. His first song to be interpolated in a Broadway show was heard in 1903. Two years later, "How'd You Like To Spoon With Me?" gave him his earliest hit. For the next nine years most of his music was heard as interpolations in other men's scores.

Then, in 1914, another interpolation, "They Didn't Believe Me," became the hit of *The Girl from Utah* and gave him a reputation he was never to lose. The success of his songs for that show earned him the offer to create a complete score for a Marie Cahill vehicle, *Ninety in the Shade*. His librettist was Guy Bolton. The 1915 musical was a quick flop, but Bolton and Kern embarked on a long friendship and collaboration.

Guy Bolton was only a few months older than Kern, having been born in England on November 23, 1884. His parents were American, however, and Guy returned home to prepare for a career as an architect. It was a career he quickly abandoned to try his luck at writing.

Mr. Popple of Ippleton was not the catchiest title for a Broadway show. Early on its title was changed to *Nobody Home*, which in 1915 had a double meaning, since in the slang of the day "nobody home" meant someone was brainless, empty of gray matter upstairs. Unfortunately, it could have served as an appropriate title for most of the brainless musical comedies inundating Broadway at the time.

Bolton has suggested that he and Kern set out to change the mindlessness of musical comedy from the very start. The show's early history belies his recollections. Bolton and Kern's original assignment was simply to "Americanize" the English show, with Bolton making only the most necessary alterations in the book and Kern interpolating a handful of songs to spice up Paul Rubens's score. When a dress rehearsal before an invited audience revealed such patching would not work, Bolton undertook a more drastic revision and all of Rubens's songs were eliminated. Kern wrote additional melodies but he did not write the entire score. In the day's dubious fashion, one that admittedly helped Kern to get his start, interpolations by other composers were brought in.

Bolton's final version of the libretto began at the Hotel Blitz (read Ritz). Violet Brinton's snobbish aunt stands in the way of her marrying Vernon Popple, "a society dancer," especially after the aunt learns Vernon has been seen around town with the Winter Garden star, Tony Miller. Tony is about to leave on a tour when Vernon's brother Freddy appears. Taken by him, Tony lets Freddy have her apartment while she is away. The action switches to the apartment, where, in tried and true musical-comedy fashion, principals enter at the most inconvenient moments and immediately misunderstand each other's relationships. Tony returns in time to clear up the complications.

Bolton's writing was full of holes. Some of the humor was outrageous and some of it hardly amusing. Cues for songs were frequently awkward. All in all, *Nobody Home* seemed as slapdash and contrived as the other musical comedies playing up and down Broadway. Comstock and Marbury undoubtedly realized this.

A shrewd barrage of publicity preceded the opening. Most likely the producers were merely trying to reassure wary playgoers who might have gotten wind of early troubles. They could not have foreseen what lay ahead. Yet their publicity brochures did make several points that proved correct, although might have been better applied to the succeeding offerings. Playgoers were promised the "smartest musical offering of the New York season," an evening of fun with "a real story and a real plot, which does not get lost during the course of the entertainment." Making a virtue out of necessity, one blurb assured potential ticket buyers, "This particular offering seems especially appropriate to an intimate playhouse of the character of the Princess."

Critics chose to judge the show on its own merits, unswayed by the ballyhoo. A majority of them liked what they saw, even if few were carried away. "A good musical entertainment with nothing of special sensational interest" was one paper's evaluation that reflected most others. Of course, there was no reason any paper should have attached any unique theatrical importance to the show. As the *Sun* observed, "very little . . . could be called important from any point of view." *Nobody's Home*'s historical importance would become obvious only with passing years. Only the *World*, among major New York papers, recognized something of the value of Kern's trailblazing song in the show, "The Magic Melody," hailing it as "quite all its title implied."

Nobody Home was only a modest hit in New York, where it chalked up 135 performances. Comstock and Marbury promptly took it on the road for several seasons. If the show's New York run had not been up to the producers' expectations, it at least showed them that musical comedies could turn a profit for the Princess. They promptly set about working on a second. This time, instead of adapting an English musical for Broadway, they elected to transform a recent American comedy hit into a song-and-dance show. The comedy chosen was Philip Bartholomae's *Over Night*. Bartholomae was signed to turn his play into a musical libretto and Kern to create the melodies. Bartholomae's story focused on the seemingly naughty, but actually quite innocent, complications that arise when the wife in one couple and the husband in another are forced to sail together on an overnight excursion. Schuyler Greene was enlisted for whatever new lyrics were needed, but since, in another of the day's peculiar practices, many of Kern's melodies were brought from earlier shows with their lyrics intact, Greene's job was not extensive.

This time around, the new musical, *Very Good Eddie*, was booked for an extended tryout to whip it into shape. A Schenectady, New York, first night revealed the show was no more ready for a New York City audience than *Nobody Home* had been at its invitational preview. The producers quickly closed it and brought it back to New York for rewriting. Rather than trust Bartholomae with the task they looked once more to Bolton. Bolton and Kern would from then on be the mainstays of all the classic Princess shows. Bolton's rewriting provided the necessary touches. His writing had generally improved. Gone were the loose ends and the irrelevant jokes. Characters actually developed during the course of the show, at least in as much as musical-comedy characters can. And Kern's songs were reasonably well integrated into the story, despite the fact that many had been culled from older shows. As a result, *Very Good Eddie* was a smash hit. The Princess shows were on their way!

One more ingredient was necessary to the alchemy that was to transmute these shows into pure theatrical gold. Even with Kern's beguiling melodies and Bolton's fast-improving librettos, something had been missing—a good lyricist. The Princess Theatre shows found more than a good one. In P. G. Wodehouse they found a great one, the finest lyricist the American musical stage had yet heard. Small matter that Wodehouse was not American.

P. G. Wodehouse was born in Guildford, Surrey, England, on October 15, 1881. After attending Dulwich College, he planned to make a career in finance but found writing more congenial. Some of his earliest English lyrics were written for Charles Frohman's London shows and had music by Kern. At the same time, he was a newspaper columnist and in 1910 published the first of his Psmith novels. Their popularity brought him to America, where he wrote for the *Saturday Evening Post* and took an aisle seat as *Vanity Fair*'s drama critic.

With the addition of Wodehouse to their roster, the Princess shows entered their all too brief golden heyday. *Have a Heart* appeared in 1916, *Oh, Boy!* and *Leave It to Jane* in 1917, and *Oh, Lady! Lady!!* in 1918. (As we have seen, an attempt at reviving the series with *Sitting Pretty* failed in 1924). For a number of reasons, neither *Have a Heart* nor *Leave It to Jane* played the Princess, although they were written to be presented there. Even the shows that did open at the Princess didn't linger there for long. They were too much in demand to be sequestered in such a tiny house where, despite the period's relatively healthy theatrical economics, it would have been financially foolhardy to keep them. *Very Good Eddie*, *Oh, Boy!* and *Oh, Lady! Lady!!* all were transferred to larger houses.

According to its first tryout reviews, *Oh, Boy!* underwent a more drastic preopening revision than either *Nobody Home* or *Very Good Eddie*. Early accounts reported it was set on or near a college campus and that its shenanigans involved collegians. Since these early reviews were highly flattering, the resetting of the story may have been prompted not by any inadequacy but by the knowledge that *Leave It to Jane* would employ a similar background. While on the road the principals were recast, slightly older players were hired, and the setting was changed to Long Island's country club world. At least one of the songs cut at this time reappeared in *Leave It to Jane*. Also dropped was a rousing, expansive chorus number, "Ain't It A Grand And A Glorious Feeling," apparently because it would have been cramped by the Princess's tiny stage.

When *Oh, Boy!* raised its curtain at the Princess, polo players, celebrating a victory, helped to set off the action. The players and their girls are smuggled into George Budd's apartment by one of the young sports, Jim Marvin. Jim's hope is to enlist George in the partying. But George is nowhere to be found, so Jim and his friends, singing "Let's Make A Night Of It," go off to continue their search and their revelry. They are no sooner gone than George appears, carrying his bride Lou Ellen across the threshold. In their newfound happiness, George and Lou Ellen cannot imagine how they survived all those years when "You Never Knew About Me (And I Never Knew About You)." Lou Ellen discovers a telegram awaiting them. It is from George's guardian, his Quaker aunt Penelope. She has learned that George is contemplating marriage and is coming to speak to him about it. George fears she will cut off his allowance when she realizes his marriage is a *fait accompli*. Lou Ellen, her eye ever on the pocketbook, offers to return home until Aunt Penelope leaves. Of course, Lou Ellen adds that she will do whatever her husband wants, since she is determined to be "An Old-Fashioned Wife."

After the newlyweds have gone their separate ways, Jim returns. He is making himself comfortable in George's living room when a young lady enters—by way of a window. She tells Jim that she is Jackie Sampson and that she is fleeing from a policeman whom she slugged during a rumpus that began when a man known as Tootles kept giving her lingering looks at an inn. At first wary of one another, Jim and Jackie are soon confessing that each could use "A Pal Like You." Jim offers Jackie the shelter of George's apartment, telling her to claim she is Mrs. Budd if anyone comes. He also offers to find Tootles. When Jim has left, Jackie changes into a pair of Lou Ellen's pajamas. George returns, and Jackie explains her problem in the nick of time, for the policeman she hit is at the door. George flourishes his marriage license, to the policeman's perplexity. Before the officer departs, he admires the pajamas, so the supposed Mrs. Budd offers to obtain the pattern for him. George would now like Jackie to leave, but since it has started raining, he politely agrees to find another bed for himself that night and allow Jackie to remain "Till The Clouds Roll By."

The second act takes place at a country club. The victory celebration is continuing. Lou Ellen and her mother arrive; so do George and Jackie. In desperation, George introduces Jackie as his Aunt Penelope. Lou Ellen and her mother are surprised, but their surprise is small compared to Jackie's when Lou Ellen's father, Judge Carter, appears. Jackie realizes he is none other than Tootles. Jackie has been open with the rest of the club guests, telling them the man she will marry must be a living catalogue of assets "Rolled Into One." Suspicious of Jackie, Mrs. Carter demands the judge keep George and Lou Ellen apart. They beg "Oh, Daddy, Please" let them sit and talk together, but the judge, under his wife's thumb, refuses. Meanwhile, Jim and Jackie find a moment alone, and recognizing that they are falling in love, dream of "Nesting Time (In Flatbush)." Lou Ellen's parents' refusal to permit her to speak to George has not daunted her, and she assures her newfound friends at the club that "Words Are Not Needed" to continue a romance. Suddenly the real Aunt Penelope appears. She is a Quaker of the old school in dress and speech. Confusion runs rampant, until the real Penelope is served a spiked drink that Jackie has ordered. When she becomes tipsy and forgiving, George clears up the confusion by confessing that he and Lou Ellen are already man and wife.

In outline, *Oh, Boy!*'s story differed little from so many other cotton-candy plots then serving musical comedies. Bolton, Wodehouse, and Kern, however, saw to it that it was something more. All of Bolton's stories were played out in a well-heeled, well-bred world, reflecting the continuing changes that were taking place both in the country and in the theatre. Increasing prosperity, particularly as the war progressed, brought with it an increasing sophistication and interest in costly pastimes. Films continued to lure away more and more of the less affluent, less educated playgoers, but as was true in Cohan's heyday, silent, black-and-white flickers could not offer serious competition to color-laden song-and-dance entertainments. Nevertheless, the theatre was slowly, subtly becoming more elitist—perhaps in the best sense of the word—and the Princess, with its meager balcony, was more elitist still.

Bolton's dialogue therefore was always stylish and knowing. More importantly, his characters were well-defined and human. Lou Ellen is his best example. She is not the saccharine, coy heroine who time and again had critics climbing walls in outrage. She is, in fact, something of a calculating bitch, a petite, slightly tempered version of her battle-ax mother. Her instant willingness to give up her honeymoon

night rather than risk jeopardizing George's allowance is an early example. Her conversation is peppered with ominous hints that she plans
to rearrange not only George's apartment (she objects to a copy of a
classic Greek nude), but George's way of life as well. Wodehouse recognized the character Bolton was developing and carefully continued
that development in his lyrics. Thus, while Lou Ellen agrees that words
alone are not needed to cement a courtship, that loving looks and
flowers are just as meaningful, she adds pointedly in a second chorus
that "diamonds or pearls on a string . . . have their language too."

Bolton also saw to it that most of the show's humor derived from
character and situation. Irrelevant Joe Miller jokes were a plague in
the period's musical-comedy librettos, but Bolton had little use for
them. In the second act the policeman confronts George, Lou Ellen,
and Jackie, whom he knows as Mrs. Budd but who is now pretending
to be the Quaker aunt for Lou Ellen's benefit:

> *Policeman.*   How about the pajama pattern?
>
> *George.*   It's coming . . . it's coming.
>
> *Lou Ellen.*   Pajama pattern?
>
> *Jackie.*   Thee shall have it, good man, I promise theee.
>
> *Policemen.*   How's that?
>
> *Jackie.*   I will even give thy good wife mine if thee will only beat
> it.

Wodehouse's lyrics were of an even higher order than Bolton's
libretto. Seldes saw them as the work of "an inventive and almost
fantastic mind," often brilliantly combining "a gentle satire . . . a
delicacy of sentiment and . . . exaggeration of sentiment." Wodehouse pins down Lou Ellen's nature in her very first chorus. After
George somewhat naïvely assures her that had they known each other
as youngsters he would have let her feed his rabbit "till the thing
became a habit," she responds by recalling kisses stolen underneath
the mistletoe by little boys "excited with tea," and she assures George:

> If I'd known that you existed,
> I'd have scratched them and resisted, Dear.

Concise comedy may have been Wodehouse's forte, but, as Seldes suggested, he was equally skillful with carefully modulated sentiment. His opening chorus of "Words Are Not Needed" is as touching as it is subtly humorous.

> If ev'ry day he reads the message he sees in her eyes,
> If, when he gazes fondly in them, she droops them and sighs.

The lines flow naturally. They capture the easy, unconsidered movement of everyday speech, free of the jarring involutions and lyrical clichés of the time. Yet Wodehouse was, while remaining natural, capable of imaginative rhymes, such as "smart is" with "Parties" in "An Old-Fashioned Wife," or of the stunning placement of a single word, such as "When it's nesting time in Flatbush we will take a little flat." There had simply been no lyricist nearly as good as Wodehouse before this.

But there had never been, and never has been, a composer to equal Kern. His songs for the Princess Theatre shows are not his best. He had yet to develop fully those free, gorgeously curvaceous lines that so identified him and gave him his greatest classics. "They Didn't Believe Me" was the only example of that high style he had produced at the time. Historians generally agree that the song established the modern ballad—the 4/4 time song—as we still know it today and allowed it to replace the waltz as the principal song in contemporary musical comedy. Only once in the Princess shows did Kern match "They Didn't Believe Me," and that is an unfortunately neglected song from *Have a Heart*, "And I Am All Alone." This exceptionally touching, plaintive ballad moves from one musical idea to the next without ever precisely repeating itself. It builds ineluctably to a searingly passionate conclusion. Yet if "And I Am All Alone" stood on a plane by itself, the rest of Kern's best songs for the Princess are marvelous indeed. "Till The Clouds Roll By," "Babes In The Wood" (from *Very Good Eddie*), and "The Siren's Song" (from *Leave It to Jane*) retain honored places in his canon, while "Bill," which most people regard as one of *Show Boat*'s gems, was first sung in *Oh, Lady! Lady!!*

Kern's difficulty at this time was that he was caught in the still rampant "dancing craze," so his songs usually had to be framed with ballroom dancers in mind. No one equaled his inventiveness in the face of these restrictions, but it was not until he liberated himself from them that his inimitable artistry found full range. One important example of that early inventiveness was "The Magic Melody," a song often credited with introducing blues or jazz harmonies to Broadway. Less obvious was Kern's clever reuse of melody within a show to set a mood and to tie the story together even more securely. In *Oh, Boy!* George writes a letter to the departed Lou Ellen to the tune they had sung at their first appearance. "You Never Knew About Me." "Oh, Daddy, Please," in which George and Ellen plead with Judge Carter, reemployed "An Old-Fashioned Wife" and "A Pal Like You."

When the Princess shows were at their height, both Bolton and Kern granted interviews in which they expressed their desire to create intelligently written, fully integrated musical comedies, devoid of the loose ends and claptrap contrivances still so rampant. Bolton boasted that *Oh, Boy!* "had nothing irrelevant in it. From start to finish it was a 'straight' and consistent comedy with the addition of music. The plot was connected, and every song and lyric contributed to the acceleration of the action. . . . The humor of 'Oh, Boy!' was based entirely on situation—not on interjected comedians." Kern echoed Bolton's sentiments, insisting that he preferred to write his melodies with specific characters and turns of plot in mind. That these pronouncements were still a bit utopian in 1917 or 1918 scarcely detracted from their essential truth. Yet the considerable rewriting of the librettos and *Oh, Boy!*'s borrowing, say, of "A Package Of Seeds" from *Ninety in the Shade* suggested that ideally integrated and rationalized musical comedy remained something of a wistful goal.

Certainly the Princess shows' competition rarely displayed Bolton, Wodehouse, and Kern's elegant flair. Even Kern's other musical comedies of the time—*Love o' Mike, Toot-Toot!, Rock-a-Bye Baby*—were, for all their contemporary professional competence, far more crudely developed. Kern's childhood neighbor, Louis Hirsch, provided scores for similar shows, notably *Going Up* and *The Rainbow Girl*. In many instances musicals were thrown together as vehicles for those "interjected comedians" Bolton professed to disdain. For example, Charlotte Greenwood toured incessantly in her Letty shows, her warm, "down-home" humor and her high-kicking dancing implanting their singular stamp on all her entertainments. The era's most popular musical-comedy clowns, Fred Stone and Dave Montgomery, moved from one personal triumph to another. Revues, of course, kept many a Broadway house lit, while operettas, although hurt by growing international animosities, still found some favor. For the most part, however, great stars, fine composers, and spectacular productions kept cash registers ringing.

Initially, the Princess shows' low budgets prevented the hiring of top stars. What began as necessity rapidly became policy. Still, so eagle-eyed were Comstock and his associates that they consistently enlisted the most promising youngsters, many of whom went on to long stardom: Fanny Brice (in a road company of *Nobody Home*), Ernest Truex, Vivienne Segal, Oscar Shaw, Edna May Oliver, and, for what it's worth, Marion Davies.

The Princess Theatre shows stand virtually alone among the era's musical comedies in still being able to enjoy successful revivals. Yet their salutory effect on succeeding musical comedies was distressingly small, and little quite like them has ever again graced Broadway. Arguments can be mustered to suggest that the better musicals of the twenties, the best Gershwin or Rodgers and Hart shows, for example, did carry over some of their improvements: the greater integration of song and story, and the more cohesive, stylish librettos. But these influences were tangential and reflected the new, pervasive artistic climate in the live theatre as much as anything. Still, the models remained for anyone wanting to look back to them.

8

Cinderella

At first, her name was Irene O'Dare, and men who sang her praises described her as "a little bit of salt and sweetness," "a dainty slip of rare completeness," and the "sort who captures hearts to charm them." Yet within less than a year some of her suitors were calling her Mary Howells and agreeing that as part of a second flattering list of charms "Nary Another girl . . . has your way with the boys." The confusion was compounded just over a month later when she was being saluted as Sally Rhinelander and assured, "There is no lady in the land that is half so fair as Sally." Yet a certain doubt had crept in by this time, for Sally's beaux sadly confessed that her clothes were "a poor affair," a far cry from what such a lovely girl should wear. In fact her beaux, a little shaken, were prepared to admit they would love her "No matter what her name." And then it dawned on them! Irene and Mary and Sally were, indeed, all aliases. Broadway's newest heartthrob was really Cinderella.

At least Broadway began to believe she was. Admittedly, she wasn't the Aschenbrötel German stories had spoken of as early as the sixteenth century, nor the Cendrillon Perrault had made famous at the close of the seventeenth century in his *Contes de ma mère l'oye*. After all, she had no wicked stepmother or wicked sisters, no fairy godmother, no pumpkin coach nor slippers of squirrel that Englishmen mistranslated as glass. Indeed, this modern-day Cinderella belonged squarely in a group that the original Cinderella would have found hard to fathom. This twentieth-century upstart was a member of the middle class. She was to be, more often than not, a pretty Irish secretary who falls in love with her boss's son. A company's president-to-be was as close as she usually came to a charming prince. Nevertheless, to Broadway she remained Cinderella, and Broadway was to embrace her as it rarely embraced anyone.

Irene and *Sally* were the biggest Broadway musical hits up to their day, and *Mary*, while not quite so successful in New York, kept several road companies touring season after season. In fact, the first of its four road companies was touring before the principal company reached New York. All three shows remained beloved long after they had taken their final curtain. The Shuberts, never shy about pouncing on someone else's success, rushed in a musical called *Sally, Irene and Mary*. It too was a hit.

The Cinderella story was not new to the musical stage. Mother Goose's Cinderella had long since been taken up by burlesque and, especially, by pantomime. In due course, the basic story—that of a poor, deprived waif who finds love and a home among the upper crust—was adapted in all sorts of guises. Operetta loved it and had flower girls and laundresses marry princes or young kings. Nor did musical comedy shun it. What would the Gaiety musicals have been had not their shop girls married into the peerage? But for three seasons—1921–22, 1922–23, 1923–24—this classic motif dominated Broadway musical comedy. Some critics feared it might monopolize it, and many a review of the period opened with a frightened shudder. "Just when we hoped we might have seen the last of her," one critic bewailed, "Cinderella has swept back into town." A hundred and twenty musicals braved Broadway in these three years. Half were either operettas or revues. Of the fifty-eight that could be branded musical comedy, no fewer than twenty-one centered on a Cinderella figure. In short, over a sixth of all musicals and well over a third of all musical comedies employed the same basic story. No wonder Broadway quickly dubbed these years its Cinderella era.

Without question *Irene* initiated the vogue. This warm, intimate musical (and it was warm and intimate in its original version!) beguiled a nation uncertain of itself after the horrors of World War I and slipping into a postwar financial depression. President Harding's assurance that he would lead the country "Back to normalcy" was still a year away. In the meanwhile, *Irene* filled delighted audiences with its cheerfulness and its charm, and its hope of goodness triumphant.

New York first welcomed *Irene* at the bandbox Vanderbilt Theatre on November 18, 1919. Its heroine was a shop girl, like the Gaiety heroines before her. When she is sent on an errand to the imposing Marshall estate on Long Island, she is innocently involved in an unpleasant misunderstanding. Luckily, the Marshalls' handsome young son Donald understands the real situation, takes pity on the young girl, and sets matters right. He also falls in love with her. To help her along, he obtains a modeling job for Irene at a chic couturier's shop run by a man known as Madame Lucy. But Donald and Irene's budding romance is opposed from two sides. Donald's snobbish mother looks down on the girl and, perhaps more surprisingly, Irene's mother is suspicious of all rich people. In the end, Irene's very obvious charms and goodness win over everyone.

Perhaps the most interesting aspect of *Irene*'s plot was Mrs. O'Dare's suspicion of the rich. It injected a relatively new note into this type of story. Class distinctions had been accepted in musicals as natural and inevitable, as a rule. Of course, there had been any number of bounders and imposters among the silk stocking set, but these characters were generally depicted as villains and lost the girl in the end, usually to a more worthy member of their class. Poor girls in operetta and earlier musical comedy customarily encountered little objection to their rising in the world. Certainly they were given little reason to mistrust the rich or titled simply as a class. In this respect then, Mrs. O'Dare's social perceptions reflected the era's growing tendency to muckrake in stories of the well-to-do.

Irene succeeded not merely because of its book. "Alice Blue Gown," the hit of Harry Tierney's score, became ragingly popular and has remained so. The rest of the score was a delight too. The bubbly title song, popular enough in its own day, was used effectively in the original production to make the point that class distinctions are meaningless if traced back far enough. It was sung by Mrs. Marshall as she desperately attempted to discover a suitable social lineage for her prospective daughter-in-law. "The Last Part Of Every Party" was an early example of the "party" songs that so accurately reflected the voguish partying of the twenties. Casting was also superior. Edith Day, the original Irene, became a major Broadway and West End star.

In *Mary*, Mary Howells was not, strictly speaking, a Cinderella figure. She had probably never known the drudgery that Irene had endured in her tenement nor the hard times Sally was to undergo as an orphan. Mary was the daughter of a college president and had taken a job as Mrs. Keene's social secretary. Mary falls in love with Mrs. Keene's enterprising son Jack. Jack, however, hardly gives Mary a second look. In an ironic twist that doesn't become evident until late in the story, the Keenes have gone bankrupt. Luckily, Jack has headed west to build cheap, portable homes he can sell to struggling white-collar youngsters. His homes don't sell all that well, but while digging a foundation on one of his properties, Jack strikes oil. He returns home very wealthy again. On his return he does give Mary a second look—and a third and fourth—so the curtain falls with everyone paired the way audiences knew they would be by the middle of the first act.

Although *Mary*'s story was not truly a Cinderella tale, contemporary critics viewed it as such. Sandwiched between the openings of *Irene* and *Sally*, its success gave further impetus to the exploding Cinderella vogue. As with *Irene*, its music as much as its story helped it along. The show-stopper was "The Love Nest," the song in which Jack describes the joys of his tiny houses (and which George Burns and Gracie Allen later made their theme song). In fact, when the show began its long tryout it was called *The House That Jack Built*. Like *Irene* before it and *Sally* afterward, *Mary* offered a captivating title song. "We'll Have A Wonderful Party" joined a growing list of twenties hedonistic anthems. The score was Louis Hirsch's finest, his lively music made all the more enjoyable by George M. Cohan's brisk, dance-packed staging.

Ziegfeld brought *Sally* into the great New Amsterdam Theatre on December 20, 1920, and consolidated beyond recall the fad for the Cinderella motif. "Nothing less than idealized musical comedy," one critic hailed the show. Because she was an orphan, Sally Rhinelander was probably the saddest waif of all. She had been given her patronymic after she had been discovered in a phone booth: the phone exchange was Rhinelander. Wealthy Mrs. Ten Broek brings Sally with a group of other orphans to the Elm Tree Alley Inn, hoping to find employment for them. A dishwasher is needed, and Sally is selected. While she is washing dishes, a patron enters the kitchen. He is a rich, handsome bachelor, none other than Blair Farquar, of the Long Island Farquars. (From show to show, the first Cinderella librettists were consistent in assigning Prince Charming's castle to Long Island.) Blair has come to arrange for a soiree. But he is sufficiently taken with the beautiful young drudge to advise her to keep her chin high—high enough, in fact, so that she can always look for a silver lining. Blair's advice so delights Sally that she dances for joy, and her dancing lands her a spot in the soiree's entertainment. The usual musical-comedy misunderstandings follow, as does the usual happy ending. Sally's dancing earns her not only a place in the *Ziegfeld Follies* but also a wedding ring from Blair.

This ending brings up an interesting point. Was stardom or marriage the ultimate prize? Ziegfeld, of course, would have chosen stardom. So, in real life, did the original Sally, Marilyn Miller, who remained Broadway's reigning musical-comedy favorite for over a decade. *Sally* was not the first show to pose the problem. For example, in Rudolf Friml and Otto Harbach's 1912 operetta, *The Firefly*, the street waif of the first act ends as both an opera star and a rich man's wife. Naturally, the question was never posed explicitly. In the fairy-tale world of musical comedy, it was simply assumed that career and marriage could go happily hand in hand.

Even more than in *Irene* and *Mary*, it was not merely *Sally*'s libretto that carried the day. From first to last, the original *Sally* displayed Ziegfeld's singular theatrical magic—the sumptuous settings of his great designer, Joseph Urban, and the artistry of his large roster of capital performers, including Leon Errol and Walter Catlett. Most of all, however, Jerome Kern's superb score gave the show its tremendous popularity. The enduring "Look For The Silver Lining" caught the facile optimism that was commonplace at the time, while "Whip-Poor-Will" may have caught with equal precision the postwar nostalgia for prewar days. Similarly, "Wild Rose," in which the heroine insists she is not "a prim and mild rose," captured something of the sweet but determined nature of the period's Cinderellas and hinted at the era's drive for untrammeled good times. These Cinderellas were not to be like the passive one of legend, who required a nudging fairy godmother and a determined prince to rise in the world. This new look in Cinderellas mirrored to some extent the sense of liberation that women enjoyed after the war (however small that liberation may now seem).

Indeed, these heroines reflected exactly the right amount of liberated behavior and attitudes, at least as certain librettists viewed it at the time. Two musicals that opened fast on *Irene's* heels implied some American girls had gone too far. In *The Rose of China* (with a Bolton-Wodehouse libretto) and in *Always You* (Oscar Hammerstein II's first Broadway libretto) pushy American girls lost out at the final curtain to more tractable foreign girls. At the same time, the classic, docile Cinderella found a small niche for herself in a successful Shubert revue, *Cinderella on Broadway.* In this show, Prince Charming's search for his lady of the slipper was merely a tenuous frame device to tie together all sorts of unrelated skits and songs. Indeed, until shortly before the revue opened, the Shuberts kept announcing that the revue would be called *Rip Van Winkle, Jr.* Somehow, their often insensitive but theatrically acute instincts told them Cinderella was the coming thing, so the title and the frame were abruptly changed at the last minute.

All of these musicals opened in 1919 or 1920, the last full years of the seven-season epoch—1914–15 to 1920–21—that witnessed the birth of the modern American musical at the Princess. They were not the only shows to utilize the Cinderella motif, merely the most successful. For example, less than a month after *Irene's* premiere, *Miss Millions* opened at the tiny Punch and Judy Theatre. Despite its somewhat deceptive title, apparently prompted by the alliteration the era so loved, the musical recounted yet another rags-to-riches tale. This time the heroine was a waitress in a tea shop. The uncle of her rich fiancé suspects her motives for marrying his nephew, so he arranges for her to think that she has come into a large inheritance at the very time her fiancé has lost all his wealth. The waitress, whose name was also Mary, remains true to her beau until she learns of the ruse. Then, believing her fiancé had no faith in her, she runs away. You can guess the end.

By the fall of 1921 Broadway's smart money had decided Cinderella meant cash in the bank, so in September of that year the parade of Cinderella musicals began in earnest. Critics quickly began to decry a certain sameness among the ladies, although they were equally quick to applaud the most deserving. These modern-day Cinderellas did, in fact, tend to share common characteristics. Alice O'Brien, Mary O'Brien, Nelly Kelly, Glory Moore, Helen McGuffey, Mary Jane McKane, and Rosie O'Reilly attested to a strong Irish strain. So did Poppy McGargle, even if she took her name from her guardian, while Jane Lee, boarding with the McGuires in their tenement, may be assumed to have been Irish as well. No fewer than seven of the girls were orphans. At least five were shop girls, while four peddled flowers on the street. One was an organ grinder, another a burglar. Several others were secretaries or stenographers. Following in Sally Rhinelander's shoes, six were performers, who usually sang or danced their way into their sweethearts' affections. The secretaries and stenographers generally had a harder row to hoe. More often than not they earned their gold band by helping the boss's son outshine his short-sighted, stubborn father, and, of course, it was that same boss's son who gave them their rings. Some of the more passive Cinderellas first called attention to themselves when they were cruelly snubbed in social situations. The injustice of the cut and the girl's dignified response won some rich boy's heart and started the romance.

Although many a waif was Irish, her rich suitor hardly ever was. Librettists often took an easy way out, by suggesting how rich or at least how socially prominent the suitors were. For patronymics they obviously went to the social register. When these girls threw off their rags for their new rich garments, two of them did so as Mrs. Harrimans and two as Mrs. Hammonds. Others left the altar as Mrs. Van Courtlandt, Mrs. Van Wick, Mrs. Morgan, Mrs. Wellington, Mrs. Winthrop, Mrs. Chattfield, and a host of similar, if less instantly recognized, names.

Yet for all the apparent and probably unavoidable similarities, librettists did manage to inject a reasonable amount of variety into these stories. Crimes of one sort or another gave a heightened theatrical coloring to several of the plots. Only the heroine of *The Hotel Mouse* was herself a criminal. She is a baby-talking gamin who robs rich hotel guests. When she makes the mistake of trying to burglarize the suite of rich Wally Gordon, Gordon catches her in the act and, instead of turning her over to the police, he sets about reforming her. Reformation and romance go hand in hand. Knowing playgoers might have expected some such story from *The Hotel Mouse*, for its author was Guy Bolton, who had often displayed a predilection for scoundrels and out-and-out crooks. Rose-Marie, the heroine of *Good Morning, Dearie*, while not a criminal herself, was the pet of a jewel thief. Only by helping to foil one of his exploits and threatening him with exposure can she and her new society lover clear their way to the altar. In a quirky twist, Cinderella's Prince Charming in a third musical, *The Chiffon Girl*, was a bootlegger.

Violence, like crime another element not in the original Cinderella tale, found a conspicuous place in several of these redactions. *Good Morning, Dearie*'s hero, Billy Van Courtlandt, first demonstrates his feeling for Rose-Marie by coming to her defense in a dancehall brawl. In *Plain Jane*, the hero, one of the many bosses' sons, quits his father's firm when his father objects to his Cinderella sweetheart and enters a prizefight contest in order to win enough money to set up a competitive firm.

In a sense, these changes or intrusions were largely decorative (to the extent that crime and violence can be seen as decorative). Certainly they did not affect the basic story. But, librettists were willing to make alterations even in the fundamental structure. Some variations were comic and superficial, some much more significant. For example, in *Just Because*, Cinderella—or Syringa, as she was rather ridiculously called in the show—is a rich girl who only pretends to be an orphan in order to catch her man. Her man, in this case, is the director of the orphanage. In one of the most successful Cinderella shows, *The Gingham Girl*, the heroine's great love is a young man who prefers to chase chorus girls and paint the town red. Rather than see red herself, the determined young lady builds a successful cookie business and is ready with money in the bank when her beau comes to his senses.

The most drastic changes were in the works of two celebrated Irishmen, George M. Cohan and Eddie Dowling. In Cohan's two noteworthy hits of this epoch, *The Rise of Rosie O'Reilly* and *Little Nellie Kelly* and in Dowling's *Sally, Irene and Mary*, the Irish heroines more or less reject the expected Cinderella ending. In *The Rise of Rosie O'Reilly*, Cohan hedged a bit. Rosie's wealthy man-about-town is disinherited by his irate father. The young man takes a job as a florist to support his bride-to-be. The father relents in time for a totally happy ending, but Cohan makes the point that Rosie would have stuck with her man through thick or thin. Neither she nor the hero was aggressive enough to establish a successful business as so many of their contemporary Cinderellas and Prince Charmings did. It was a touch of hard-headed realism in a fairy-tale world. Cohan had been even more realistic a year earlier in *Little Nellie Kelly*. Nellie is a policeman's daughter who is courted by two quite different men. Jack Lloyd is polished, wealthy, and in the social register. Jerry Conroy is brash, wisecracking, and struggling. Although Jack courts Nellie with lavish parties in elegant mansions, Nellie chooses Jerry. (Insiders might have known in advance that Nellie would marry Jerry, for Nellie and Jerry were the names of Cohan's parents. But Cohan's plot was clearly not determined by family ties.) Dowling's *Sally, Irene and Mary*, which had begun life as a vaudeville spoof of the three earlier hits, had a similar ending. Mary, the central figure, throws in her lot not with the socially select Rodman Jones, but with Jimmie Dugan, a rising Irish politician.

There was hardly any contest when it came to deciding just where Cinderella's beautiful promised land was. It was right here in the United States, and, if the settings of most shows were accepted, in New York

City at that. Of the two dozen Cinderella musicals produced at the three-season height of the craze, all but a handful unfolded not far from the banks of the Hudson or of Long Island Sound. One was set in Italy, two in France, and one on an ocean liner headed for America. A fifth began in the alleys of New York but ended on the stage of a West End theatre.

Musically, the shows were something of a mixed bag, ranging from melodies that bordered on operetta (and were occasionally perceived as such by some critics) to the snappiest latest styles influenced by jazz. Ivan Caryll in *The Hotel Mouse*, Rudolf Friml in *Cinders*, Herbert Stothart and, surprisingly, Vincent Youmans in *Wildflower* represented the older school trying to be modern, while J. Fred Coots in *Sally, Irene and Mary*, Harry Tierney in *Glory*, and, not surprisingly, Vincent Youmans in *Lollipop* and *Mary Jane McKane* (the latter again with Stothart) spoke more assertively in up-to-date terms. Jerome Kern employed his unique idiom in *Sally* and *Good Morning, Dearie*.

Dancing was an important part of all these musicals, and, for the most part, was still the "go-into-your-dance-after-your-song" variety. Isolated productions occasionally offered decorative ballets, but the ballets scarcely advanced the story. Indeed, more often than not, the story stopped cold to permit the dance spectacle. *Sally*'s gorgeous Butterfly Ballet (with music by Victor Herbert, not Kern) may have shown how Sally Rhinelander triumphed in the *Ziegfeld Follies* and thus had a tenuous connection with the plot, but it was really there to allow Ziegfeld and Urban an eye-filling production number. Its choreography was light and classical. Very few shows followed *Sally*'s example. For the most part, dancing consisted of soft-shoe routines, ballroom dances, and tap numbers. There was one minor change. Precision dancing (as opposed to the earlier precision marching) became the rage at this time. Many critics felt these numbers were more gymnastic than choreographic. Long lines of beautiful girls interspersed their tapping with all sort of bends and high kicks and somersaults. But whatever its nature, stepping became such a conspicuous part of these musicals that many were looked upon as essentially dancing shows.

Why Cinderella? Why, for that matter, Cinderella in modern dress?

Why so many Irish Cinderellas, so many Cinderellas in business, so many Cinderellas as stage stars? And why the deviations from a traditional tale that has stood the test of time so well and for so long?

Musicals have always reflected their times, as noted earlier. The Cinderella musicals that so preoccupied the early 1920s were no exception. The Princess Theatre shows had typified and crystallized an earlier school with elitist settings and elitist stories. But from the time films had lured away large numbers of blue-collar audiences, the theatre itself had grown increasingly elitist, so librettists had responded accordingly. This explains the waning popularity of George M. Cohan's written-for-the-man-in-the-street musicals, even when the characters professed to be upper class. World War I had shaken society, not as drastically as World War II was to do, but it had brought about perceptible cracks in society's former impregnability. Taxation, however small by later standards, had also begun to chip away at the upper class. Then came Prohibition, which allowed many who had believed themselves shut out from normal business channels to find a means to quick, if illicit, wealth. A disproportionate number of these profiteers were Irish, Italians, Jews. These and other minority groups had long found a genuine welcome in the arts, and nowhere more than in the theatre, but rarely had they been accepted in the mainstream of society. Now, in illegal and legal enterprises, minorities began to weave themselves into the national fabric.

Of all the minorities, the Irish were most like the more Anglo-Saxon majority, but for their ardent Catholicism. It was only natural that sooner or later they would begin to find an important place not only in society, but in reflections of that society on the musical stage. Admittedly, those reflections were rosily colored, but that had always been the way of the theatrical world. While the Irish comprised a sizable segment of playgoers, there remained substantial anti-Irish prejudice, which may account for the deviate endings both Cohan and Dowling offered. Cohan and Dowling were known to have special followings among Irish who otherwise rarely patronized the theatre. These were often less affluent Irish, less educated, less polished, less sophisticated, and so less likely to believe they stood a realistic chance of rising in storybook fashion.

Like the Irish, Jews made up a sizable share of theatrical audiences, although their numbers may not have been sufficient to allow a show to run on their patronage alone—especially outside of New York. Strong prejudice against Jews remained throughout the American heartland. As a result, librettists shied away from Jewish heroines. That many Jewish entrepreneurs had fared handsomely in business, that their sons and successors well might fall in love with a Jewish girl from one of the many Jewish tenements apparently were facts shrugged off as promising small theatrical returns at best. This reluctance to use Jewish principals was shared by Jewish producers, writers, and composers. Jews remained largely the butts of jokes on the musical-comedy stage.

Significantly, *The Chiffon Girl,* the only musical about Italian-Americans (*Wildflower* dealt with Italians in Italy), gave the hand of its Italian-American heroine to an Italian-American bootlegger. Audiences still might have balked at an Italian marrying into American society. Curiously, several earlier operettas had allowed just that. But *Naughty Marietta* and *The Firefly* had been written for a tiny, beguiling Italian, Emma Trentini, whose charms and glorious singing obviously overrode any objections to her nationality.

A rage for black musicals exploded all along Broadway at the time of the Cinderella craze. None of these resorted to the Cinderella motif. There was no question at the time of a black waif finding a white Prince Charming. That was not only undesirable, but in many states actually illegal. Yet a black Prince Charming was also out of the question, for the tremendous disparity in status that was a basic part of the Cinderella story did not apply to blacks. Virtually all blacks were poor, and the few exceptions were hardly upper class. So black musicals relied on the same comic stereotypes that had been their stock-in-trade since the days of *Clorindy* and Bert Williams and on the uniquely zestful black dancing that appealed to this age of "dancing shows."

Racial and ethnic bigotry pervaded these musicals—no doubt reflecting the same hardening of attitudes that allowed an organization like the Ku Klux Klan to flourish. Terms such as "coon," "nigger," "wop," "yid," "chink," and even "mick" abounded not only in many a musical's dialogue, but even in their unpublished stage directions.

So deep did these feelings run that Irishmen could be figures of ridicule in the very same shows that boasted Irish heroines. Harrigan's compassion and Cohan's somewhat preachy tolerance were things of the past.

Calvin Coolidge's comment that "the business of America is business" was still a few years in the future, although Ulysses S. Grant or William McKinley might have made it years before without raising many eyebrows. In the early twenties the perception was widespread, and not even a brief postwar recession could dampen America's growing admiration for the business world. If the rich were still very rich despite taxation, the middle class was probably more affluent than ever, even the lower middle class. And it had found new freedoms in the wake of the war. It was not uncommon for secretaries who still worked a five and a half day week to leave their offices at noon on Saturday, have lunch, and then head for a matinee. They no longer had to be escorted or chaperoned. They brought with them to the second balcony a sense of business that their bosses carried into the more expensive seats. There was precious little chance that they would rise to executive positions in that day and age, but they could catch an executive's eye and heart. Most would marry within their own set, but they could dream, and the theatre helped to shape that dream for them. Since getting to the top was often more fun than being there, stories of business success may have held more interest than tales of society dalliance, which, in any case, had come to seem a bit threadbare. An even more glamorous route was the stage. Besides, stage stories offered all sorts of opportunities for intra-theatrical fun. Of course, marriage remained the be-all and end-all of most girls' dreams, so stardom in these musicals went hand and hand with romance.

The intrusion of a criminal motif in some of these stories reflected America's growing urbanization, since crime was perceived as being far more ubiquitous and organized in big cities than elsewhere. Except for the burglaring heroine, the crimes and the criminals clearly were viewed as city types and primarily Italian. The bootlegger in *The Chiffon Girl* was an Italian, as was the gangster from whom *Good Morning, Dearie*'s Rose-Marie eventually breaks loose.

By the middle twenties, a surfeit of Cinderella shows had set in.

Of course, only the critics had sat through them all and early on had complained about their similarity. So Cinderella shows might well have continued to appear left and right for season after season, despite critical howls. But in the fall of 1924 *Rose-Marie* and *The Student Prince* became Broadway's darlings and revived a vogue for operetta. Coincidentally, a new style of musical comedy emerged. Together the new shows changed Broadway's ideas of what musical comedy should be. Although Cinderella was not to be totally forgotten, she was relegated to her ashes.

9

All That Jazz

In 1922 the famous blackface team of McIntyre and Heath opened at New York's Shubert Theatre in a seemingly harmless little musical called *Red Pepper*. Heath played a cocksure opportunist called Jimpson Weed, the "Get-Rich-Quick Wallingford of the Colored Race," while McIntyre portrayed Weed's hard-luck buddy, Juniper Berry, a "Colored Gentleman of Misfortune." Weed believes their chance for quick riches lies with a bedraggled horse named Red Pepper, and their attempt to cash in on the nag's hoped-for winnings takes the pair all over the map. By the show's end they have succeeded in making a number of other people rich, but they are as broke as they were at start. In the fashion of many blackface plots, the pair wander off at the final curtain to seek easy money elsewhere.

Critics were not overwhelmed by the story or by the dialogue. Most felt it was typical of the era's blackface foolery, with all the expected blackface routines. They were not offended by the stereotypes, as some would be today. Indeed, they were sure there were many potential patrons who would enjoy the show. Most were essentially noncommital about the music—most but not all. The critic for the *Post* moaned in high dudgeon, "there's nothing in the piece that is real music. Jazz, yes. Any amount of it, but jazz isn't music. Not by a long shot."

Attacks on jazz were widespread and growing. John Philip Sousa wrote slightingly of it, assuring his readers that both Beethoven and good march music would outlive this passing fancy. No less a figure than Jerome Kern issued several broadsides. Kern had been among the first to introduce many jazz harmonies in his Broadway songs and jazz instrumentations as well. Saxophones were first heard in a theatre pit band at Kern's musical comedy, *Oh, I Say!* Kern's rejection of jazz was not a blanket rejection, yet his displeasure touched on what for many is the very heart of jazz—improvisation. Kern felt improvisation went too far, destroying carefully worked out tempos and harmonies and playing havoc with equally thoughtful melodic lines. Early in 1924 he bravely put his money where his mouth was. He refused to allow public performances of his music from his latest show, *Sitting Pretty*. The score was one of his finest, but the public silence prevented its becoming popular and the show failed.

With knowledgeable artists railing against the new musical form, it was hardly surprising to hear many other voices raised against it. Much of the protest was the sort of disparagement that besets any novelty. And to no small extent, jazz was not yet totally understood. Misunderstanding aggravated the shock of newness. Social prejudices also played their part. With time, however, the newness wore off and understanding spread. At least as far as Broadway was concerned, the date when jazz became an established and even welcome musical idiom can be pinpointed to December 1, 1924. On that night *Lady, Be Good!* opened. Musical comedy was never quite the same again.

To comprehend fully just what change *Lady, Be Good!* brought about we need to backtrack a bit. For several seasons Cinderella musicals and revues had dominated Broadway's musical stages. Operetta had taken a back seat since the war, despite the successes of *Maytime*, *Apple Blossoms*, and a very few other, more or less traditional examples. Americans seemed to prefer lighter, more obviously native musical styles. Some shows after the war offered songs considered jazz numbers, but critics were quick to dismiss them as a passing fad. When *Blossom Time* opened in September 1921, Stark Young began his review in the *New York Times*, "After jazz, what? They tried a new answer on Broadway last evening. . . ." That answer was, of course, revived operetta. *Blossom Time* became one of the 1920s' biggest hits. But for three years the show stood in grand isolation. It neither revived operetta nor displaced jazz. (How interesting that so perceptive a critic as Stark Young, while not dismissing jazz out of hand, implied it was little more than an ephemeral vogue.)

The musical that pushed aside Cinderella shows and truly restored the popularity of operetta was *Rose-Marie*. This Rudolf Friml-Herbert Stothart-Otto Harbach-Oscar Hammerstein II opus opened in September 1924 and quickly became the most successful musical of the 1920s. It ran a year and a half in New York; road companies abounded. Its London success far surpassed its Broadway stand, while in Paris, where it ran for three years, it established a long-run record. The night after *Lady, Be Good!* arrived on Broadway, *The Student Prince* premiered. Its triumph, coming just three months after *Rose-Marie*'s, clinched the new rage for traditional operetta. And traditional it was! With earnest romanticism and exotic settings these operettas jumped back to the days before *The Merry Widow* to present totally old-fashioned, if wonderful, escapist evenings. For the next five years Broadway luxuriated in the last magnificent heydey of the genre.

On February 12, 1924, almost a year prior to *Lady, Be Good!*'s unveiling, George Gershwin, with strong assists from Paul Whiteman and Ferde Grofé, had brought jazz permanently into the concert hall and had thereby given it a cachet and respectability it had lacked. While there had been a few earlier attempts to give jazz a concert-hall imprimatur, it remained for Whiteman's *Experiment in Modern Music* to secure the seal of approval. The recital that February evening at New York's Aeolian Concert Hall had included works by Victor Herbert (who died three months later) and Irving Berlin, but the program's momentous breakthrough had come with the playing of Gershwin's "Rhapsody In Blue."

When *Lady, Be Good!* opened at the end of the year with its jazzy Gershwin score, it presented critics and playgoers with another distinct kind of book musical. They now could choose between otherworldly operettas offering soaringly lyric, European-influenced music and up-to-the-minute musical comedies, enlivened by less grandiose, patently American musical mannerisms. Distinctions between the two types of musicals, which often had been so fuzzy, were now sharply drawn— and rarely would be fudged or confused again (although, as we'll see, Gershwin essayed a few "political operettas" in the idiom).

Gershwin's music was not pure jazz. No commercial songs could be considered pure jazz. The essence of jazz was its attitude toward a given melody as a starting point, its need for improvisation more than any peculiar melodic lines, special harmonies or tempos. Nevertheless, jazz did employ peculiar melodic lines, special harmonies and tempos, and it was those characteristics that Gershwin appropriated and developed in the songs he wrote with his brother Ira. The curvilinear lines of operetta (and of the music of such independent-minded masters as Kern) gave way to far more angular melodic lines. The classical harmonies were interspersed with fresh, narrow, "bluesy" modulations, while the more gently flowing tempos of the older schools were replaced by more staccato, excited tempos, as well as by the distinctive languor of pure blues. The resulting sound was steely, often sardonic, and always thoroughly contemporary.

Indeed, "thoroughly contemporary" described every aspect of *Lady, Be Good!*, from its sleek, brightly colored Norman Bel Geddes sets with their hints of the growing art deco fashion to the glitteringly polished, lean, vaguely debunking mannerisms of Fred and Adele Astaire and their excellent supporting players. Guy Bolton and Fred Thompson's story, set in a world of expensive Pierce Arrows, cloche hats, and bootleg champagne, caught the era's flippant yet assertive optimism. That story centered on Dick and Susie Trevor who have been dispossessed by Dick's vengeful ex-girlfriend. Their brash, wise-cracking lawyer, who by musical-comedy coincidence is also the ex-girlfriend's attorney, attempts a reconciliation. In exchange for his advice, he requires Susie to disguise herself as a Mexican lady. The inevitable complications set in quickly as do the inevitable romances with their happy endings.

But for all the brilliance of Bel Geddes, the Astaires, and their associates, it was Gershwin who carried the day and whose songs have allowed the show to endure. Curiously, the songs were not well integrated into the show. Several had hackneyed and contrived lead-ins, of the sort that had long plagued musical comedy. The touching parting sentiments of "So Am I" were trivialized when the lovers' parting turned out to be only until breakfast. Clearly, the lesson of the Princess Theatre shows—that songs could be as artfully integrated into musical comedy as they often had been into operetta—had been forgotten.

Jacob Gershvin was born in a Brooklyn tenement on September 26, 1898, less than two years after his brother Israel (Ira) and just one day before Vincent Youmans. Father Gershvin was a ne'er-do-well businessman, but his commercial failures never soured the Gershvins' home life. In 1910 the Gershvins bought a piano, ostensibly for Ira. George took to it immediately; Ira preferred books. Under the tutelage of Charles Hambitzer, a pianist and composer, George received a thorough classical training. His affections, however, lay with the musical ferment all about him. At the wedding of an aunt he was struck by some songs the band was playing. He learned they were Jerome Kern's interpolations for *The Girl from Utah*. From then on, Kern was a shining exemplar to the neophyte.

In 1919 Al Jolson introduced the Gershwin-Irving Caesar song, "Swanee," in *Sinbad*, and it was an overnight success. Most of Gershwin's better writings were heard in revues for the next five years, although he also composed scores for several book musicals. Out of these early shows came "I'll Build A Stairway To Paradise" and "Somebody Loves Me," songs which clearly brandished Gershwin's original signature. Out of *Lady, Be Good!* came three Gershwin classics and three almost canonized songs: "Oh, Lady Be Good!," "Fascinating Rhythm," "So Am I," "Hang On To Me," "The Half Of It, Dearie, Blues," and "Little Jazz Bird."

During the rest of the decade, Gershwin followed up with a succession of memorable scores. "Looking For A Boy," "When Do We Dance?," "These Charming People," "That Certain Feeling," and "Sweet And Low-Down" came from *Tip-Toes*. "Maybe," "Clap Yo' Hands," "Do, Do, Do," and "Someone To Watch Over Me" came out of *Oh, Kay!* Audiences left *Funny Face* humming not only the title song but "He Loves And She Loves" and " 'S Wonderful." Even his failed shows left wonderful songs behind. *Treasure Girl* introduced "I've Got A Crush On You"; *Show Girl*, "Liza." (With collaborators Gershwin also worked on at least one operetta, *Song of the Flame*, and one curious hybrid, *Rosalie.*)

Cataloguing this long list of songs underscores a major point about jazz-age musicals: their songs set them apart (and in the long run it seems it is only their songs that endure, for few of the shows are ever revived). In these five years or so, Gershwin more than any other composer seemed to speak to the world with a totally American voice. Of course, a song is not merely a melody. A song is both words and music, and Gershwin was blessed with an outstanding lyricist, his brother Ira. The newly liberated musical comedy of the 1920s introduced Broadway to some of its greatest lyrics writers: Ira Gershwin, Lorenz Hart, and Cole Porter. Writing in free and easy colloquial styles, they brought a grace and wit to the Broadway lyric that only P. G. Wodehouse before them had matched.

Ira's lyrics were the sunniest, rarely trafficking in Hart's consistently misanthropic philosophy or Porter's silky, sometimes arch, sophistication and his obsessive sexual innuendoes. Perhaps nothing came to represent Ira's fresh approach more than his ability to catch America's popular vocal shortcuts. " 'S wonderful! 'S marvelous," one of his most famous refrains begins. In another he captured the slangy black patois with "Clap-a-yo' hand! Slap-a-yo' thigh!" Even when employing more everyday speech, he caught its natural flow. The hero of *Lady, Be Good!*, alarmed that the heroine might marry someone else, cries, "Life would be the bunk/I'd become a monk." When the man of Tip-Toe's dream first came into her life, she confesses she grew "dizzy;/Thought, 'Gee, who *is* he?' " Time and again Ira's rhymes were felicitously impudent, pairing "enjoy it" with "Detroit," and "faux pas" with "blokes pass." But his sense of fun was not confined to wordplay. He could sustain an idea just as cleverly. Witness the tale of Mr. Smythe-Smythe in "These Charming People," when "some twenty-seven" children turn up for the reading of the childless man's will.

Magnificent melodies burst forth all over Broadway in the middle and late twenties, and they were increasingly accompanied by superior lyrics. One thing, however, did not change very radically. Librettos remained much the same, the sort of sugar-coated nonsense that audiences loved (and still do) but that more pretentious critics then and now have looked on condescendingly. Cinderella scarcely was remembered, although she was not left behind entirely. In the librettos for Gershwin's shows, several of the heroines may have had holes in their stockings at the beginning of the evening. But no one dwelled on their financial plight (although *Lady, Be Good!* opened with a newly dispossessed brother and sister hanging their "God Bless Our Home" sign on a lamppost).

By 1925 American prosperity was so rampant that even arrivistes were pretending they had arrived long ago. As a result, tenement and factory settings, standbys of the Cinderella era, all but disappeared. In their place were the watering holes of the rich—country clubs, luxury hotels, ballrooms, and mansions. Ragamuffin dresses and business suits gave way to jodhpurs, plus-fours, evening gowns, and top hat, white tie, and tails. The good life as the twenties perceived it, the life of free-flowing, illegal hooch, flat-chested flappers, all-night parties, touring cars, and the Charleston, was the heady milieu of these musicals. This may not have been the world of the very rich, of old money and stand-offish society, but it reflected the ways of those who were rich enough and who were often willing to join wholeheartedly in the slumming and capering of the more showy nouveau riche.

One plot will have to suffice as an example. Paul Gerard Smith and Fred Thompson's story for Gershwin's 1927 hit, *Funny Face*, moved from rich men's homes to a fashionable resort inn to a suite at a luxury hotel in Atlantic City to the Two Million Dollar Pier (the famous Million Dollar Pier was really there). Even the name of the hotel, the Paymore (a send-up on the real Traymore) spoke of this era's preoccupation with money and moneyed people. In the story a rambunctious hoyden named Frankie bristles under the strict guardianship of Jimmie Reeve, who locks her jewels in his safe and places obstacles in the way of her romance with Peter Thurston. Frankie and Peter plot to steal the jewels. So do two comic crooks, Herbert and Dugsie. Love triumphs after two hours of complications, singing, and dancing.

Plots and dialogue were not that original in these shows. Broadway's infatuation with the pun, or at least with wordplay, had scarcely diminished. Dugsie informs Herbert, "I'm an artist. I etch." Herbert responds, "Then why don't you scratch yourself?" In mirroring the smart set's brash cynicism, a certain knowing bitterness crept into the dialogue. It had often been evident in earlier shows, but it became a far more fashionable way of thinking and speaking at this time. A character in *Lady, Be Good!* warns his girlfriend, "I don't want to reform too darn fast. It might make you lose interest if there wasn't something about me for you to pick on."

Styles of humor change rapidly, and much of the humor in these shows seemed atavistic even in their own day. But with such capital funny men as Victor Moore, William Kent, and Walter Catlett delivering the lines, audiences were glad to be tolerant. And then there were the Astaires, and, in *Oh, Kay!*, Gertrude Lawrence, adding their own unique luster.

Gershwin's musicals might be seen as the cream (or champagne) of contemporary musical comedies; those by the team of De Sylva, Brown, and Henderson were its meat and potatoes. They were more popular and better reflected the real world of the tired businessmen and their wives who were then, as now, a mainstay in the box office. Like Gershwin before them, most of De Sylva, Brown, and Henderson's earliest writing was for revues. But when they turned their hands to book shows, they depicted not the glittering world of high society and slick social climbers, but the more or less everyday bourgeois world around them—theatrical people, happy-go-lucky college kids, prizefighters, and amateur golfers. While their songs were far less imaginative or stylish, they were instantly likable, perhaps more than Gershwin's on first hearing. So well did their songs catch the pulse of their day that their 1927 musical, *Good News*, more than once has been called the quintessential musical comedy of the 1920s.

Superficially, *Good News*, was just another tale of collegiate romance and football that followed in the wake of *Leave It to Jane*. " 'Good News,' " Brooks Atkinson reported in his morning-after notice for the *Times*, "rah-rahed, joked, flirted and finally made an exciting end-run and a touchdown for 'dear old (fill in name of)'—in this instance, Tait College." But at football-mad Tait, Tom Marlowe, its team star, will not be allowed to play if he fails his astronomy exam. He passes the exam after he is carefully tutored by Connie Lane, a pretty coed who loves him but believes he prefers Patricia Bingham. Tom almost loses the game when he fumbles, but clownish Bobby Randall recovers the ball, saving the day for everyone. The victory celebration takes an added meaning for Connie when she realizes Tom does not love Patricia.

The story, even in 1927 becoming old hat, was merely a frame on which to hang the high jinks that so characterized the "era of wonderful nonsense." For example, the show opened with the orchestra, all in collegiate blazers, running down the aisles and cheering as they took their places to launch into the overture. And the songs that overture included! "Good News," "Just Imagine," "Lucky in Love," "Varsity Drag," and "The Best Things In Life Are Free" are still lovingly sung, hummed, and whistled by graying men and women. The love songs are unashamedly melodic and romantic, while "Varsity Drag" and even "Lucky In Love" are irresistibly toe-tapping. They have none of Gershwin's daring inventiveness nor his more outreaching harmonies. Indeed, there is a four-square simplicity about most of them. And certainly their lyrics lack Ira's freshness and impishness. But they capture the essence of most of the period's popular tunes. And much the same may be said of De Sylva, Brown, and Henderson's other shows and scores for this time: *Manhattan Mary, Hold Everything!, Follow Thru*.

Since an almost incredible array of brilliant, adventuresome young composers (and a few brilliant, adventuresome young lyricists) accounted for much of this era's greatness—as well as for virtually all of its durability—several other memorable figures must be touched on, however briefly. Vincent Youmans's *No, No, Nanette* was the 1920s' most successful musical comedy and, after *Rose-Marie*, Broadway's most popular export. Like the librettos for Gershwin's shows, Otto Harbach and Frank Mandel's saga recounted the dalliances of the rich. In this case, a tightly leashed young lady and her free-spending, free-thinking, Bible-publishing guardian and uncle are the central figures. The lady longs for love and the carefree sprees she sees her flapper friends enjoying. She runs away to Atlantic City at the very time her guardian is there meeting some of his own young lady friends. Both the heroine and her uncle are pursued by her irate aunt. Youmans's score offered not merely the perennial "I Want To Be Happy," but the most popular song to come out of the era, the consummate soft-shoe number, "Tea For Two." His remaining songs, all instantly likable and whistleable, ranged from somber blues to pulsating tap-dance melodies. A superb cast was assembled for the original New York production, led by vivacious, dimple-kneed Louise Groody and dapper, droll Charles Winninger.

In 1926, Youmans's less successful *Oh, Please!*, another story of society peccadilloes, left behind "I Know That You Know." A year later, Youmans set to music a story more on the order of those in De Sylva, Brown, and Henderson's shows. *Hit the Deck!* followed a group of rowdy sailors and their sweethearts on a trip around the globe. "Hallelujah!" and "Sometimes I'm Happy" came from its score. Youmans's musical-comedy style was extraordinarily spare. For many of his best numbers he relied on economic, closely knit phrases played over fresh, changing harmonies and occasionally altered in tempo. After *Hit the Deck!* he spent the rest of the decade trying unsuccessfully to write operettas.

Time may well show that the best musicals of the period were those written by Richard Rodgers and Lorenz Hart, often to librettos by Herbert Fields. Indeed, the trio combined successfully so often that for a short time they were looked upon as apparent heirs to Kern-Bolton-Wodehouse. Many of Rodgers and Hart's shows of these years are especially interesting because they hint at Rodgers ultimate drift into operetta. While most contemporary musical comedies continued to be set in the "here and now," *Dearest Enemy*, *A Connecticut Yankee*, and *Chee-Chee* all unfolded long ago or in faraway settings, or both. Even the dream milieu of *Peggy-Ann* might be seen to contain seeds of operetta otherworldliness.

Although Fields's librettos were highly praised when their shows opened, today they seem to differ little from other good librettos of the time. Those who like their sort of light-hearted, slightly light-headed manner will find them eminently workable; those who don't will dismiss them as dated. But initially and enduringly the songs beguile. Rodgers's style, while slightly angular in keeping with modern tastes, manifested his debt to and admiration of his idol, Kern. Like Kern's music, his was sweeter than most, exhibiting that carefully restrained charm, that "odor of sachet" that was Kern's trademark. Luckier than Kern at this time, Rodgers had an incomparable lyricist. Hart had all of Ira Gershwin's virtues: easy flow and lyrics that were rarely stilted or involuted. He had Ira's gift for impudent rhymes. Many a writer has pointed to Hart's rhyming "go to Hell for ya" with "Philadelphia" in his very first Broadway show, although they have ignored his matching "I'm goin' to corner ya" with "California" in the same number. Years later he paired "trousseau" with "Robinson Crusoe." Hart lines also often had a more literate air about them, perhaps because Rodger's musical lines sometimes seemed longer and more curvaceous than Gershwin's. His lyrical sentences are often surprisingly long, filled with delightful internal rhymes: "There's a dash in it/Of a passionate tender feeling, too." But most of all, Hart had a distinctive view of life. Admittedly, it was a smilingly jaundiced view, for deep within Hart was a tortured soul. His light touch, however, made his views as acceptable as they were biting and pervasive. Writing of female virtue, for example, he noted, "But the purest driven snow will sometimes drift, you know." "A little birdie told me so!" he added hastily, not mentioning the bird was, like him, probably a gloomy night owl.

A *Connecticut Yankee* is typical of the trio. Like the Mark Twain story on which it was based, Fields's libretto had fun with anachronistic juxtapositions of modern and medieval matter, although a number of critics complained that Field's joking injection of archaic words grew tiring. Too many "wouldsts" and "thous" soon palled. The story began at a modern bachelor party, on the eve of the hero's wedding to a bitchy young lady. Knocked unconscious during the revelry the hero dreams he is at King Arthur's court. He brings about all sorts of modern improvements until the machinations of Merlin and Morgan Le Fay (played by the same actress who portrayed his bride-to-be) make things too hot for him. On regaining consciousness, he realizes that he does not love his fiancée, but another girl. Rodgers and Hart's capital songs include two that became standards, "My Heart Stood

Still" and "Thou Swell." John F. Hawkins's wittily expressionistic sets added color to the festivities, while Busby Berkeley's dances enlivened them.

Toward the end of this era, Cole Porter began to come to the fore. Porter had been represented professionally on Broadway long before Gershwin, Henderson, Youmans, or Rodgers, but his first efforts had met a discouraging reception. He had then spent several years in Europe. He returned at the end of the twenties and immediately enjoyed two successes, *Paris* (whose score included "Let's Do It") and *Fifty Million Frenchmen* (which offered "You Do Something To Me"). These shows and their scores mirrored Porter's own penthouse world. *Paris* was a relatively intimate, one-set vehicle designed around the talent of Irene Bordoni, who played a young French lady in pursuit of a rich, handsome New Englander. In one novel departure, Irving Aaronson and his ten-piece jazz band, the Commanders, were placed onstage instead of in the orchestra pit. *Fifty Million Frenchmen* was far more typical of the era's big, lavish musical entertainments. Its plot centered on a rich man who bets he can win a girl without flaunting his wealth. He wins both the bet and the girl.

Of the older composers, Kern remained in the vanguard, although the mid-twenties were unkind to him. Apart from *Sunny*, with its unforgettable title song and "Who?," his scores produced no classics, and a surprising number of his musical comedies were quick flops. *Sunny* was created as a vehicle for the most dazzling of all 1920s musical-comedy queens, Marilyn Miller, although its producer, Charles Dillingham, filled his cast with other superior talents, including Jack Donahue, May Hay, Clifton Webb, Joseph Cawthorn, and "Ukulele Ike" Edwards. A persistent legend has it that Dillingham instructed Otto Harbach and Oscar Hammerstein II, his librettists, to create a story around specific settings—a circus, a fox hunt, an ocean liner, and a grand ball. The simple tale they concocted of a circus girl who must choose between two suitors dutifully included all of them. Yet so loose was their story that in New York Sunny chose one suitor, in London the other. Not until 1927, when Kern turned to operetta with *Show Boat*, did he revive his creative juices.

There were many other fine composers, a few good lyricists and librettists at work on Broadway in the last half of the twenties, but to catalogue them here would be unreasonable, for the list would run on and on. One other aspect of these musical comedies, however, must be treated. Unfortunately, it is the most difficult, for it is so ephemeral. Glamorous mountings and, even more importantly, glamorous stars accounted for so much of Broadway's greatness.

Physically, jazz-age musical comedies presented a mixed bag. In an interview in 1979 Vivienne Segal told me that if I were to see some of the productions she had starred in during the twenties I would "double over with howling" at their tackiness. Shubert productions were frequently assailed by the critics, for the penny-pinching brothers were wont to reemploy old sets and costumes wherever possible. On the other hand, many producers coupled an understanding largesse with superb taste. Ziegfeld, of course, remains legendary. His house designer, Joseph Urban, was possibly the greatest theatrical decorator in our history. He was supreme both as a stylist and as a colorist. No man was more influential in ridding musicals of the tasteless kaleidoscope of colors that had splashed across every scene in older musicals. Urban carefully controlled his colors, not only limiting them to shades of a single color within a scene, but frequently seeing to it that they flowed gently from one scene into the next. Robert Edmund Jones, James Reynolds, John Harkrider, and a host of less celebrated artists also brought their taste and imagination to the musical stage.

When these men designed costumes, they had their work cut out for them in more ways than one. Since the stage was still on a firm footing economically, choruses remained large. Forty, fifty, or sixty members continued to be common, while some choruses approached a hundred.

The dancers who stepped out of this crowd continued to employ time-tested routines. Dances changed little through the twenties and for years thereafter. Waltzes, tangos, taps, Charlestons, and precision drills remained standard, while ballet, when it was employed at all, was stubbornly traditional and flowery, except in an occasional progressive revue. Critics frequently complained about a certain sameness but in the very next line often confessed the dancers' infectious vitality had eventually won them over.

But songwriters aside, the glory of American jazz-age musical theatre was its galaxy of shimmering stars. Until sound films lured many of the best entertainers to Hollywood, Broadway offered a variety of great performers unmatched before or since. Nowhere was this more in evidence than among the clowns. The Marx Brothers, Ed Wynn, Fred Stone (after Montgomery's death), Clark and McCullough, W. C. Fields, Leon Errol, and Beatrice Lillie were such capital funmakers that their names above titles were often enough to ensure profitable runs. In fact, the shows they appeared in were usually vehicles written especially for their unique talents and were often worthless without them. When a superior writer handed over material to them, he did so at his own peril. George S. Kaufman was delighted to hear the Marx Brothers actually recite one of the lines he had written for them in *The Cocoanuts,* while Irving Berlin's delightful score for the same show was lost in the laughter. For the most part, these great comedians used the basic material given them simply as a starting point. Their extended visual routines, their inspired improvisations were their mainstays. Their comedy ranged from Stone's innocent gymnastics through Ed Wynn's outrageously naïve lunacies (with his lisp and weird costumes) through Clark and McCullough's more suggestive but good-natured romping through Field's misogynistic inebriations and Errol's wobblier ones through Miss Lillie's unique blend of hauteur and hellion to the hilariously destructive mockery of the Marx Brothers.

Two singing comedians were top attractions. In fact, Al Jolson was probably Broadway's hottest ticket through much of this period, although he was the first to leave for Hollywood. Performing in blackface, his brash style and uniquely warm singing overcame the barrier his burnt cork created. Jolson often set aside the show and sang for his audiences long into the night. It was difficult to determine whether a loosely structured show such as *Big Boy* was a true musical comedy, a glorified revue, or simply a unique Jolson vehicle. By contrast, several of Eddie Cantor's last Broadway musicals were indisputably book shows. *Kid Boots* and *Whoopee* were among the decade's biggest musical-comedy successes. Unlike Jolson, who performed against legendarily shabby Shubert backgrounds, Cantor's jittery exuberance competed with Ziegfeld's incomparable opulence. Cantor's boyish enthusiasm, his jumpy, hand-waving delivery, and his famous eye-rolling often won the moment. He sometimes performed in blackface but never to the extent Jolson did. Nor did he ever dominate a whole show the way Jolson could. Aging playgoers, looking back over the years, can remember much about *Kid Boots* and *Whoopee* besides Cantor, while few can remember anything about Jolson's or the other great comics' shows except Jolson's singing and the others' clowning. Shows that starred less side-splitting performers seem more memorable in the long run, suggesting that somehow hilarity is more blinding, monopolizes more attention than the most scintillating glamour.

The most glamorous star in the 1920s was unquestionably Marilyn Miller. Aside from a brief, unsuccessful fling as Peter Pan, she enjoyed nothing but triumphs throughout the decade. Three shows kept her busy for nearly ten years: Kern's *Sally* and *Sunny* and finally Gershwin and Romberg's half-operetta, half-musical comedy, *Rosalie*. The sprite-like Miss Miller has often been compared to a delicate blonde Dresden figurine. On stage she conveyed an impression of daintiness if not fragility, whatever her sometimes sordid private life might have been. She was a graceful, albeit unimaginative, dancer, and her voice, in those pre-microphone, pre-amplification days, rarely elicited complaints. Just how good an actress she was is moot. Some critics approved of her acting; some didn't. In common with so many leading ladies, she occasionally had to accept charges that she was too wooden or too coy. Other attractive young ladies, such as Mary Eaton, Queenie Smith, and Louise Groody, sang and danced their way into audiences' hearts, but they sang and danced in Miss Miller's shadow.

Those stars who did come to rival Miss Miller as the decade moved on were of a different ilk. Their stock-in-trade was hardly sweetness and a stagey ingenueness. They conveyed an aura of blasé cynicism and a certain world-weary hardness. England's Gertrude Lawrence and Fred and Adele Astaire were leading exemplars of this new breed. Miss Lawrence was something of an oddity in musical comedy since she was at best a mediocre dancer and she sang off-key. But she had an indescribable stage presence that overrode her obvious faults. Adele and Fred Astaire, like Miss Lawrence, had "a way" with a song, although neither of them was a particularly strong-voiced singer. Nor was either especially good-looking. Adele was considered the more talented half of the brother-sister team. She was a consummate high comedienne. More than one critic put her in a class with Beatrice Lillie. And, of course, the Astaires could dance! Their dancing was not only stylish and lively, it was often humorous. Unlike Miss Miller's careless, gay tripping, the Astaires' dancing seemed to impart something of their voguish sophistication with its tongue-in-cheek outlook on life.

Great stars and great songwriters gave musical comedy a golden age, one which reflected the insouciant, high-living era. It was too good to last. Hollywood discovered sound, and Wall Street discovered it had laid an egg. Musical comedy in the thirties was to be a markedly different form.

10

The Depression

The 1930s and its Depression were not a totally dark age for musical comedy, but the Great White Way glittered less brightly than it had in the gaudy, giddy twenties. Even if Broadway had exploded with light, there would have been fewer habitués there to enjoy it. Playgoers who just a season or two before had thought nothing of plunking down twelve dollars for a pair of tickets to the latest musical were now desperately counting pennies to see if they had enough money for groceries. Backers who cavalierly had given over thousands of dollars to mount new shows were declaring themselves bankrupt. So were many of the greatest producers—Arthur Hammerstein, Charles Dillingham, even the Shuberts. As audiences melted away and fewer shows were produced, theatres also began to disappear. Many went over to films, often as grind movie houses, or to radio. Some simply sat dark. A few became parking lots.

One curious result was that the legitimate theatre probably became more elitist in these years than it ever had been or was to be again. Live plays were mounted for the few who remained affluent and for a small band of dedicated loyalists willing to scrape together the price of a ticket. This was a haphazard elite, not bound by any unique claim of lineage, old wealth, education, or political persuasion. It brought with it to the theatre no special theatrical philosophy. For the most part, like the mass of lost playgoers, it asked only to be entertained.

In response, Broadway, however reduced its circumstances, otherwise changed very little. Disillusionment with the older order of things, endemic to some degree at almost all levels, gave rise to a number of protest plays. With remarkable speed, protest even found a voice in the period's musicals. However, with minor exceptions, most notably in a handful of left-wing revues, protest was far less shrill or fiery in musicals than it was in contemporary drama. Musicals were still designed to divert the tired businessman and his wife. John Mason Brown might write in 1938 that "musicals are now produced on the assumption that our business men and women are less tired than they used to be," but writers and producers of song-and-dance entertainment recognized that any change in attitude was one of degree, not kind.

The very first musical to arrive in the new decade underscored this point. *Strike Up the Band* brought a breath of fresh air to our musical stage. It may not, strictly speaking, have been musical comedy. For all its recourse to jazz idioms, it was closer to operetta—"political operetta," its lyricist, Ira Gershwin, called it. Bygone musical comedies often had been sprinkled with sharp satirical thrusts, but as a rule biting moments were few and far between. As often as not they were confined to lovers' quarrels or a comic's fleeting observation. The rest of the time, an almost simpering sweetness prevailed. In *Strike Up the Band* George S. Kaufman and Ira Gershwin wove a consistent, cynical, if sunny, point of view into the entire fabric of their work. Actually, Kaufman and the Gershwins had written *Strike Up the Band* in 1927, and the show they had offered Philadelphia in the fall of that year was a sharper, more piercing satire than the one they ventured forth with three years later. But the hedonistic twenties had been too busy luxuriating in good times to pay much attention. The show had been withdrawn before it reached New York.

Kaufman's story was a spoof of militarism and big business. An American cheese tycoon persuades the government to go to war with little Switzerland over a tariff on cheese. When the tycoon is exposed for using poor milk in his cheese, he turns into an ardent pacifist. The war fever, however, is out of control. America, having won the war after decoding Swiss yodeling signals, then prepares to battle Russia over a caviar tariff. The version that Broadway saw in 1930 was a softened one, for Morrie Ryskind had joined with Kaufman in removing some of the most cutting jabs. Even the basic story was altered, the cheese giving way to confectionary chocolate, and the whole yarn was portrayed as a dream. These changes in the text and a changed attitude among playgoers combined to turn *Strike Up the Band* into a hit. The team followed with two more "political operettas"—*Of Thee I Sing* in 1932 and *Let 'Em Eat Cake* in 1933.

Success or failure, all three shows left behind songs which have since become Gershwin standards, songs which out of their theatrical context sound far more like excerpts from musical comedy than from operetta. *Strike Up the Band* offered "Soon," "I've Got A Crush On You," and its title song; *Of Thee I Sing* also gave us a title song as well as "Who Cares?" and "Love Is Sweeping The Country"; *Let 'Em Eat Cake* contributed "Mine."

Despite *Strike Up the Band*'s success—its six-month run was no mean achievement for the time—Broadway did not fall all over itself to follow in its trailblazing. For the rest of 1930 the few musical comedies to reach Broadway looked remarkably like many musicals that Broadway had applauded or booed all through the twenties. A modern-day Rip Van Winkle mixed up with some none-too-villainous bootleggers (*Ripples*) gave way to a droll newspaper vendor's fairy-tale fantasies (*Simple Simon*). Boy met girl in a story about flying (appropriately called *Flying High*), then girl rescued old flame from a potentially disastrous marriage (*Jonica*)—the modern miss packed a revolver and actually fired it. And so, on and on. One musical was simply a threadbare tour of Paris (*Hello, Paris*). Even Cinderella, played by an aging Marilyn Miller, popped up briefly (*Smiles*). Not until December did the new look finally reach musical comedy with a show that billed itself as "A Sociological Musical Satire." Imagine a 1920s musical billed that way!

The show was *The New Yorkers*. Its bilious Herbert Fields libretto

took an often cold look at what one of its Cole Porter lyrics called a
"dear old dirty town." Fields's heroine was a boozy young socialite
who is head over heels in love with a cold-blooded killer. Her parents
proudly parade their extracurricular romances along Park Avenue,
where, in the evening's most-quoted line, "bad women walk good
dogs." Of course, this was still 1930 musical comedy, so Jimmy
Durante mugged as a hit man whose victims always come disconcert-
ingly back to life. Near the end the gangster-lover-hero is sent to jail
not for his killings, but for parking too near a fire hydrant. Fields
framed the whole story as a bad dream, as Ryskind had with *Strike Up
the Band*. If Porter's music failed to catch the story's misanthropic
stance as consistently as Gershwin's had caught *Strike Up the Band*'s
razor-sharp mockery, it was, nevertheless, not inappropriate. His lyrics, world-weary, literate, and probing, were what any play doctor could
have ordered. Indeed, his lyric to "Love For Sale" offended many a
critic. Charles Darnton of the *Evening World* snarled that it was "in
the worst possible taste." It was, for the time, and was banned from
radio. Porter, perhaps more than any other lyricist, was opening the
door on any number of hitherto taboo subjects, and, though he continued to have occasional detractors, he delighted many by doing it.
"Let's Fly Away" also became a standard, as did a song Porter added
during the run, "I Happen To Like New York."

Most of the musicals that followed *The New Yorkers* had their detractors too. Not because they violated long-standing taboos, quite the
opposite. Most used time-tested formulas, only this time the formulas
flunked the test. Collegiate hoopla, more boy-meets-girl—in Hollywood or in gangsterland—even a "bunion derby" with no one less
than W. C. Fields in charge: they all flopped.

And yet, one by one, musical comedies started to appear that attempted to be a little different by making some sort of small, substan-
tive statement. Audiences glancing at the program for *Free for All*
might have suspected something was up. There was no chorus listed,
no line of high-kicking, attractive youngsters with painted-on smiles.
When the curtain rose, audiences were in for a second surprise. *Free
for All*'s story—another boy-meets-girl saga at heart—took serious looks
at communist recruiting, psychoanalysis, and free love. Unfortu-

nately, Oscar Hammerstein II, who collaborated on the book and wrote all the lyrics, lacked a sustained satiric touch. The show ran only two weeks.

Nikki was a venturesome piece that attempted to rediscover the recently Lost Generation. It also killed off most of the principal men in its cast, a no-no in musical comedy. Fay Wray, who played the titular heroine, was luckier here than in films. Her final embrace was not in a gorilla's arms but in the arms of the man who later went to Hollywood as Cary Grant. *Everybody's Welcome* left its audiences in doubt as to whether its boy and girl, who lived together, were husband and wife. *Here Goes the Bride* took a nonchalant look at divorce. A black musical comedy called *Sugar Hill* offered a more realistic view of the ghetto than any offered previously, along with stereotypical comics and dances. Only *Everybody's Welcome* made the grade—and not by much.

A number of eyebrows were raised by musical comedy's exploring hitherto untouchable subjects, but only funnybones were tickled when Broadway saw *Face the Music*. Yet *Face the Music*'s libretto was as trenchant in its own way as those of *Strike Up the Band* and *Of Thee I Sing*. In one way it was more so: the Seabury Investigation of political corruption in New York City was still in full swing, and Moss Hart's imaginary tale of similar chicanery often named names, however gingerly. The leading characters were fictitious, but oh, how true to life they were! In simplest terms, Hart's story told how the police launder their graft by backing a revue by a Ziegfeld-like producer. (Ziegfeld's shows sometimes had been bankrolled by well-known mobsters.) When the new show seems on the verge of closing for lack of business, its fate is turned around by the police insisting that the producer add some smut. In musical-comedy fashion, the story had a happy ending, but by that time Hart had hilariously made some telling points.

Since *Face the Music* was musical comedy, it had to have songs. That assignment fell to Irving Berlin. Critics and public alike were delighted with them. "Let's Have Another Cup O' Coffee" and "Soft Lights And Sweet Music" have long since entered Berlin's canon of classics. To be strict about it, a sense of tonal consistency really didn't call for "sweet music." Moreover, his lyrics, while colloquial and snappy, were no match for Hart's stylish innuendo and mordant wit. Berlin's forte, for better or worse, was directness to the point of over-simplification. *Face the Music* was only a modest hit, chalking up a run of twenty-one weeks.

Topical satire did not necessarily mean a devilish treatment of venality. It could have fun with private idiosyncrasy and public lunacy as well. Many of the later shows that elected to treat this area of social foibles had longer runs which may have reflected the improving economic conditions of society rather than any public distaste for corruption or barbarity. Moss Hart's libretto for *Jubilee* had fun at the expense of England's royal family; his collaboration with Kaufman, *I'd Rather Be Right*, toyed with the head of America's royal family, F.D.R. himself, in a spoof of the New Deal. Perhaps because names were not merely named in this case but actually brought to life on stage, the story, as in *Strike Up the Band* and *The New Yorkers*, was framed within a dream. At the very end of the decade, Sam and Bella Spewack twitted high-level diplomacy in *Leave It to Me!* Porter provided the songs for *Jubilee* and *Leave It to Me!*; Rodgers and Hart for *I'd Rather Be Right*. Since none of these shows maintained a totally consistent satiric attitude, allotting several scenes to standard boy-meets-girl musical-comedy moments, it mattered little that the music was a mixed bag. When the situation called for it, Porter's and Hart's lyrics rose to the occasion. Yet it remains telling that *Jubilee*'s "Begin The Beguine" and *Leave It to Me!*'s "My Heart Belongs To Daddy" are these shows' best remembered songs.

I'd Rather Be Right and *Leave It to Me!* both ran nine months, the longest stands of any of the 1930s topical musical comedies. Nevertheless, neither show wound up in the top ten on the list of the decade's long-run musicals. They were, in fact, thirteenth and fourteenth. Not bad, considering that 175 musicals reached Broadway during those ten years, but not exceptional. The two longest runs of the period were both revues and both oddities. Olsen and Johnson's *Hellzapoppin* was a brainless, zany vaudeville; *Pins and Needles* a left-wing lark. With 1,404 and 1,108 performances respectively, they were the only musicals to run over five hundred performances. It should be noted, however, that *Pins and Needles* played at what once had been the Princess Theatre, a house about a fifth the size of a normal Broadway theatre, so its long run may not be totally indicative of simply its popularity. Well below these two champs were another revue, three operettas, and six musical comedies: *Of Thee I Sing, Anything Goes, DuBarry Was a Lady, As Thousands Cheer, The Cat and the Fiddle, Flying High, Music in the Air, I Married an Angel, On Your Toes,* and *Roberta.*

The list suggests that Broadway's surviving playgoers, however much they were willing to accept more thought-provoking entertainments, still really preferred escapist fare. Luckily for these theatre buffs, many of the best creative artists remained loyal to live theatre. The only major new composer listed in the credits for the longest-run musicals was Harold Rome, who wrote the songs for *Pins and Needles.* Otherwise, it was Jerome Kern, Cole Porter, Richard Rodgers, George Gershwin, Irving Berlin, and Ray Henderson who sent audiences away whistling, much as they had during the twenties.

Of course, none of these men, except the late-arriving Porter, was as active on Broadway as he had been the decade before. *Face the Music* was Berlin's only book show. Kern offered four shows, only two of which were musical comedies, *Roberta* and *Very Warm for May.* *Roberta* succeeded despite a deadening libretto about love in a chic Parisian couturier shop. Perceived by many critics as little more than a glorified fashion show, the musical was given a magnificent mounting and a superb cast that included old-timer Fay Templeton, Lyda Roberti, Tamara, Ray Middleton, Bob Hope, and George Murphy. Observers agreed, however, that Kern's brilliant score snatched the musical from the jaws of failure, a score that offered "Yesterdays," "The Touch Of Your Hand," "You're Devastating," and, above all, "Smoke Gets In Your Eyes." *Very Warm for May,* set on the straw-hat circuit, failed despite a score topped by "All The Things You Are."

Aside from "political operettas" and *Porgy and Bess,* Gershwin also offered two musical comedies: *Girl Crazy,* a hit, and *Pardon My English,* a flop. *Girl Crazy's* playboy hero hires a very New Yorkish cab driver to take him out west, where he hopes to find peace and contentment. Instead, he finds excitement and romance. He transforms a dude ranch into a casino, makes the cabbie a sheriff—the cabbie speaks to the Indians in Yiddish—and falls for the town's postmistress. Allen Kearns, Ginger Rogers, and Willie Howard played the principals, but a performer in the relatively small role of the saloon keeper's daughter stole the show. She was, of course, Ethel Merman, and her showstopping rendition of "I Got Rhythm" capped her sensational debut. Gershwin's memorable score also included "Embraceable You." Lucky audiences heard very special sound emanating from the pit, for the orchestra included the yet unrecognized talents of Glenn Miller, Gene Krupa, and Benny Goodman.

Arthur Schwartz, who began to make a name for himself in 1929 revues, continued to compose primarily within that form, but he also gave Broadway *Between the Devil* and *Stars in Your Eyes* as well as two musicals that straddled the fence between musical comedy and operetta. But the most productive musical-comedy composers of the 1930s were unquestionably Cole Porter and Richard Rodgers. Their careers at this time are not only interesting, but in retrospect hint at their future development.

Porter was born with the proverbial silver spoon in Peru, Indiana, on June 9, 1891. Coddled and encouraged by his mother, he began music lessons at an early age and published his first compositions when he was only eleven. To please his father he entered Yale with the intention of studying for a career in law, but by the time he reached Harvard Law School his college songs had become so popular that even the dean advised him to change his plans. Broadway heard its first Porter songs in 1915, but they were received largely with indifference. He did not return to musical comedy until the late twenties.

Porter did the scores for eight musical comedies: *The New Yorkers, Gay Divorce, Anything Goes, Jubilee, Red, Hot and Blue!, You Never Know, Leave It to Me!,* and *DuBarry Was a Lady*. They represent, in varying degrees, the personal dichotomy that was Porter: the elegantly polished brahmin who relished a brash, sometimes raunchy, change of pace. *Gay Divorce* and *You Never Know* (although the latter was overproduced) were essentially drawing-room musical comedies with principals in evening gowns or black ties and a liberal flow of champagne. *Jubilee* attempted to catch this same world at its highest level but off its guard. *The New Yorkers*, as we have seen, portrayed this upper crust at its most unsavory, mingling willingly with the most vicious criminals. Even *Leave It to Me!* dealt with a moneyed crowd, although the money in this case was very green indeed and had to be spread around to buy power. That money buys a reluctant, mild-mannered businessman, who is pushed by his ambitious wife, the ambassadorship to Russia. With Victor Moore as the businessman and Sophie Tucker as his wife, the show was never mistaken for a serious, sociological essay.

Another newly rich lady, seeking to pave her way to the top with dollar bills, was the heroine of *Red, Hot and Blue!* Superficially, the musical resembled *Of Thee I Sing*, but in wit and consistency of tone it belonged to another world. The millionairess elects to hold a giant lottery for charity. The winner will be the person who discovers the whereabouts of a girl who was branded after she accidentally sat on a hot waffle iron. Senators buy tickets hoping to pay off the national debt. An inmate of a swank prison is released to help. He goes reluctantly, since he has just been promoted to captain of the prison polo team. Then the Supreme Court intervenes, declaring the lottery unconstitutional since it benefits the American people. Unlike *Of Thee I Sing, Red, Hot and Blue!* was at heart another love story with all the standard musical-comedy trimmings. Ethel Merman, Jimmy Durante, and Bob Hope were the principals, thus setting the tone for comedy.

Anything Goes also had a comic criminal as a principal figure, but he was a wistful, mooning one rather than the raucous clowns *Red, Hot and Blue!* had employed. The Reverend Dr. Moon was "Public Enemy #13," determined but bumbling in his efforts to rise to the top of the F.B.I.'s list. He wanders aimlessly through the entertainment, generating laughs but having little effect on the principal story. That story begins when Reno Sweeney, an evangelist turned bar hostess, admits to Billy Crocker how much she likes him. A heroine telling a leading man she likes him was nothing new, but Porter gave his heroine a socko opening when her confession allowed "I Get A Kick Out Of You." Neither Reno's guileless sentiments nor Porter's daring and skill moves Billy, whose heart is set on Hope Harcourt. But then Hope claims she does not give a hoot for Billy and sails away to marriage with a titled Englishman. Billy boards the ship to dissuade her, so Reno adds her name to the passenger list. Since one confession deserves another, Billy tells Hope that "All Through The Night" he thinks only of her. Hope professes she is still unsympathetic. That leads Billy to give Reno a second look. They tell each other "You're The Top." Reno is delighted but uncertain how deep Billy's feelings run. After all, at a time when even the best writers use only four-letter words, "Anything Goes." The captain of the ship, remembering giddy times before the Depression, attempts to rouse his passengers with an old-fashioned revival meeting, and Reno, remembering her earlier calling, helps the session along by pleading, "Blow, Gabriel, Blow." When the ship docks in England, Hope learns she is an heiress. She no longer needs the Englishman's money. She reconsiders her rejection of Billy, so Reno turns for attention to the rejected Englishman. William Gaxton was Billy; Victor Moore, Dr. Moon; and Ethel Merman, Reno. The show also featured striking art-deco sets, including a triple-tiered view of the ocean liner's decks and a pert, smartly tapping chorus.

Anything Goes's characters were drawn as much from the middle class as from the upper brackets. Apart from some references to J. Edgar Hoover, his F.B.I., and the celebrated criminals of the day, its libretto took little interest in topicality. It was brash and contemporary and, as such, presaged the nature of musical comedy for the next decade. Indeed, *Anything Goes* has been called the quintessential musical comedy of the 1930s, even more frequently than *Good News* has been said to represent the twenties.

Certainly Porter's 1939 musical, *DuBarry Was a Lady*, exemplified this style still more assertively if not as memorably. Its story told of the forlorn love of a nightclub washroom attendant for a nightclub singer and of his dream that he was Louis XV and she Madame DuBarry. In this case the dream, like the nightclub settings, allowed for lavish production numbers and was not meant to soften a tough story.

Porter's songs for all these shows were Porter's songs. They never gave the impression of having been written specifically to further a plot, to maintain a carefully contrived tone, or to create a unique mood. Many seemed like they could have been employed inter-changeably, and many of them were not written for the shows in which they were finally used. Reusing songs was a common practice at the time, but Porter's melodies and lyrics have an unusually detached quality to them. More so than most great composers' songs, Porter's songs seem to be in his shows because he deigned to put them there, not because they necessarily belong there. And yet what songs they often were! Does it really matter that "Love For Sale" was originally heard in *The New Yorkers*, that "Night And Day" and "After You, Who?" came from *Gay Divorce*, or that "All Through The Night," "Blow, Gabriel, Blow," "Anything Goes," "I Get A Kick Out Of You," and "You're The Top" were first applauded in *Anything Goes?* For the record, *Jubilee* gave us "Begin The Beguine" and "Just One Of Those Things"; *Red, Hot and Blue!* offered "Down In The Depths (On The 90th Floor)," "It's De-Lovely," and "Ridin' High"; *Leave It to Me!*, "My Heart Belongs To Daddy"; while *DuBarry Was A Lady* included "Do I Love You?" and "Friendship."

Porter's lyrics went hand in glove with his melodies, juxtaposing polished, svelte lines with the most outlandish informality and some-times testing the limits of civilized naughtiness. *Jubilee*'s "Just One Of Those Things" offers the famous line, "A trip to the moon on gossa-mer wings," as well as this far less known couplet:

> As Columbus announced when he knew he was bounced,
> "It was swell, Isabelle, swell."

This unique meshing of jocular absurdity and formalized elegance made Porter a cavalier among lyricists. Yet for all Porter's virtues, his lyrics and his attitude toward using songs were the most superficial of our great songwriters. Musically and verbally, Rodgers and Hart probed deeper. They wove their words and melodies into songs that might seem as silky as Porter's but were actually far tougher-fibered.

Richard Rodgers was born in New York on June 18, 1902. Al-though his father was a physician, the Rodgers home was musical and young Richard was given piano lessons at an early age. A dozen or so visits to hear Kern's music in *Very Good Eddie* played a large role in determining his choice of careers. While at Columbia Rodgers wrote music for college shows, and his lyricists there included Oscar Hammerstein II and Lorenz Hart.

Hart was seven years older than Rodgers, having been born in New York on May 2, 1895. He liked to claim the German poet, Heinrich Heine, as a distant ancestor. While still in grammar school he was taken to his first play and thereafter became a compulsive theatregoer. Hart was educated at select private schools before entering Columbia. He left without graduating to translate plays for the Shuberts.

Together Rodgers and Hart wrote the songs for nine musical comedies in the 1930s: *Simple Simon, America's Sweetheart, Jumbo, On Your Toes, Babes in Arms, I'd Rather Be Right, I Married an Angel, The Boys from Syracuse,* and *Too Many Girls.* Although only *The Boys from Syracuse* eschewed a contemporary setting, deriving from *A Comedy of Errors* and unfolding as it did in ancient times, many of the pair's other shows continued to give hints of operetta's unworldliness. *Simple Simon,* like *Peggy-Ann* before it, dealt with dream worlds—in this case Ed Wynn's preposterous fairy-tale dreams. The singular, colorful fantasy of circus life provided the background for *Jumbo,* while fantasy of a different sort propelled *I Married an Angel.* The whole of their modern political fable, *I'd Rather Be Right,* was couched as a dream. Even the backstage milieu of *On Your Toes* might have been perceived as imparting a special, slightly exotic glamour, although backstage yarns long had been a musical-comedy standby.

On Your Toes's libretto, based on an idea by Rodgers and Hart themselves, was complicated, not very original, and not well motivated. It told of the son of old vaudevillians who has been pushed into becoming a professor of music but longs to be a simple hoofer. When he mounts a jazz ballet, his leading man, pursued by gangsters, disappears, so he is forced to assume the role himself. If the basic story was not very new, *On Your Toes* was in one respect an innovative show, for it brought contemporary ballet into modern musical comedy. Ballet in musical comedy had been largely ornamental. But the ballet that the hero mounts in *On Your Toes,* "Slaughter On Tenth Avenue," was woven, however loosely, into the fabric of the story at the point when the gangsters come to shoot the hero, mistaking him for the original leading man, and the girl, whom the hero ultimately marries, warns him in time. George Balanchine choreographed the ballet. Rodgers's superb score was capped by his music for "Slaughter On Tenth Avenue" but also included his durable "There's A Small Hotel." Ray Bolger starred.

Though they gingerly touched on operetta's purview, these musicals were essentially very good musical comedy indeed. From these entertainments, besides "There's A Small Hotel," came such evergreens as "My Romance," "Where Or When," "My Funny Valentine," "Spring Is Here," and "Falling In Love With Love."

A careful study of these Rodgers and Hart songs might uncover some telltale developments. But both men came to their art with their geniuses in full flourish, and to all but the most discerning student their early songs seem every bit as wonderful as their later ones. Even Rodgers's sometimes surprising modulations, which grew more daring and offbeat after he collaborated with Hammerstein, were present almost from the start of his career, as were his graceful musical lines. Nor did Hart's seemingly easy way with words and his dyspeptic view of human nature alter much over the years.

The thirties witnessed an evolution in mountings. No doubt goaded by Hollywood's competition, Broadway would no longer accept the simplicity and, in many instances, the tackiness that had characterized set designs, and, to a much lesser extent, costumes. The trend in costuming went almost to the other extreme. The lavish, sometimes grotesque, costumes that chorus girls and stars strutted about in increasingly were perceived as coarse and old hat. Costumes became streamlined even when they weren't always simplified. The new look on stage—so often sleek art deco in style—was enhanced by lighting which began to grow truly artful in this period. Footlights quickly disappeared, and stages were lit more effectively from lights hung on the front of balconies or elsewhere in the auditorium.

To some extent set designers, costumers, and even lighting designers had their work made easier for them by another change. Stages were no longer quite as cluttered with people as they long had been. The lure of Hollywood and radio, the demands of the ever more powerful unions, and the exigencies of the box office discouraged producers from employing mammoth choruses and large casts. This change took place very slowly, and most playgoers, perhaps most producers, were not fully aware it was happening.

Dancing changed too. While all the older, accepted routines continued to be dependable standbys, modern ballet began to gain a foothold. Playgoers who were not ballet aficionados had received their first taste of modern dance in revues. Musical comedy had adopted the new art thanks in no small measure to Rodgers and Hart, and their librettists. In 1936 *On Your Toes* had introduced "Slaughter On Tenth Avenue," and Rodgers and Hart's 1938 *I Married an Angel* offered two modern ballets because Vera Zorina was one of their stars. But these ballets were still basically ornamental. They were not integrated into the plot by modern standards. The power of "Slaughter On Tenth Avenue" came from its vibrant music, its intriguing story, and the great dancing of Ray Bolger and his associates. The ballet's connection with the principal story of *On Your Toes* was tenuous at best.

In some respects the musical comedies of the period took little heed of fashions surrounding them, a sure sign that Broadway felt it was catering to an aging audience. Nowhere is this more apparent than in the failure of swing to pervade musical comedy as ragtime and jazz had done earlier. Tellingly, almost all the musicals that took full advantage of the new style were black, and still there were qualifications. Two of these musicals were merely jivey versions of Gilbert and Sullivan: *The Swing Mikado* and *The Hot Mikado*. The third was a musicalization of *A Midsummer Night's Dream* called *Swingin' the Dream*.

Broadway's stars changed too during this period. Perhaps the most striking change was in leading ladies, exemplified by the replacement of the porcelain, dainty Marilyn Miller with the brash, brassy Ethel Merman as the musical theatre's reigning queen. Miss Merman was scarcely the decorative beauty Miss Miller had been, but she had other attractions. Her tough, New York accent, which she never totally lost, branded her as a hometown girl. Her singing voice, hardly lyrical, was superbly placed and enhanced by impeccable diction. Rightly, she never essayed the "sweet young thing" roles Miss Miller and earlier musical-comedy stars had trafficked in. There was always to be a touch, sometimes more than a touch, of low-life hardness in the roles she accepted.

Clowning became less physical. William Gaxton (the romantic lead in A *Connecticut Yankee* and *Fifty Million Frenchmen*) joined with the beloved old-timer, Victor Moore, to form a new team, yet, though they played together regularly, they rarely interacted directly with each other the way more tightly knit teams had. Gaxton's sharpness, nevertheless, contrasted effectively with Moore's wispy, bumbling but sometimes hard-nosed antics. Jimmy Durante, separated from his old vaudeville and nightclub routines, found success with his knockabout fun-making and fractured English. A few clowns of the older school remained, notably Bobby Clark, Bea Lillie, and Bert Lahr.

When Astaire was drafted by Hollywood, Ray Bolger became Broadway's favorite dancer, his sly, seemingly lazy dancing a marked contrast to Astaire's brittle fast-stepping.

All in all, then, musical comedy evolved gradually but steadily throughout the decade. Hitherto taboo subjects were treated openly and with relative honesty, mountings improved, and so did dancing. The shows remained brash, brassy, and contemporary, and theatregoers were apparently satisfied. Yet, whether they wanted it or not, revolution was just around the corner.

11

An Experimental Season

By and large, the most interesting and most successful American musical comedies—often one and the same—disappeared without establishing major trends. *Adonis* had appeared almost at the end of burlesque's heyday. Harrigan and Cohan had no real heirs. The influences of the Princess Theatre shows was small and uncertain. *Lady, Be Good!*, which established the jazz musical and marked for the first time an incontestable distinction between operetta and musical comedy, came as close as any to being a landmark. Curiously, the 1940–41 Broadway season raised the curtain on two musical comedies that might well have proven milestones but didn't. For good measure, that same season brought forth a third intriguing work which in its own way might have been something of a trend-setter. But it too was not.

By modern musical-theatre terms, the 1940–41 season was odd. Only eleven new musicals opened, the lowest figure in decades. Moreover, the openings were bunched together, the last coming at the end of January, when the season still had five months to run.

The season opened with a bang in mid-September when Al Jolson returned from years in Hollywood. His vehicle, *Hold On to Your Hats*, was an immediate hit and played to packed houses for twenty weeks, until Jolson (not his audiences) grew tired. The show's closing marked Jolson's last appearance on the Broadway he had once dominated. A few weeks after Jolson arrived, another beloved old-timer, Ed Wynn, found success with a revue, *Boys and Girls Together*. It was his next-to-last Broadway appearance. The parade of popular favorites winding up their careers continued when Joe Cook clowned in an ice revue. Three more revues also tried their luck, though only one, *Meet the People*, found some favor. The Shuberts trucked in a creaking operetta, and trucked it out again after just a single week. Improving economic times allowed Ethel Merman and Cole Porter to turn *Panama Hattie* into the first book musical since the twenties to run over five hundred performances. Slick, gaudy, and brassy, the musical told of a bar girl who wins the affection of a socialite by first winning the affection of his little daughter. Porter's score was hardly first drawer, although "Make It Another Old-Fashioned, Please" is still heard occasionally. In short, it was a generally insipid season but for the remaining three musical comedies: *Cabin in the Sky*, *Pal Joey*, and *Lady in the Dark*.

Cabin in the Sky appeared first. If it was the least successful of the trio, its failure to run as long as the other two probably must be attributed to the fact that it was perceived as a "colored" show. Until recently, except for *Shuffle Along*, black musicals never enjoyed the long runs that rewarded great white shows. Yet apart from its story and cast, *Cabin in the Sky* was a very white show, indeed—to no small extent White Russian. Vernon Duke composed the score, George Balanchine directed and, with Katherine Dunham, staged the dances, while a third Russian, Boris Aronson, created the sets and costumes. "Thick-as-borscht" Russian accents prevailed at rehearsals. The libretto and lyrics were also the work of talented whites, albeit Yankees. Lynn Root wrote the story, young John Latouche the rhymes. Significantly, the musical was Root's brainchild, and he had a completed libretto ready before a note of music was put to paper. By starting with a singularly strong book, *Cabin in the Sky* aligned itself firmly with *Pal Joey* and *Lady in the Dark*.

The musical was distinctly offbeat. More than one student has seen in its story an intriguing amalgam of *Liliom*-like fantasy and the black folklore of *Green Pastures*. Its pious, long-suffering heroine, Petunia Jackson, prays fervently that God will forgive her thoughtless, hot-headed husband, "Little Joe." God hears her prayers and gives Joe six months in which to reform. He even sends a heavenly emissary to help Joe. Not to be outdone, Lucifer, Jr., rushes up to see that Joe sticks to his mean ways. For a while, it appears that Joe really may turn over a new leaf. Then he and Petunia get into an argument, and Joe shoots her. They come to Pearly Gates where Petunia's fervent pleas for compassion move God again, so he allows Joe to enter with Petunia.

Duke's melodic score eschewed Negro rhythm and blues, yet poignantly and sensitively captured the nature of both the characters and the story. His two best numbers, the title song and "Taking A Chance On Love," quickly entered his canon of classics. Even the secondary numbers, notably "Honey In The Honeycomb," were superior.

" 'Cabin in the Sky,' " the *Times*'s Brooks Atkinson wrote, "ranks with the best work on the American musical stage." There were, of course, sterotypical elements in its treatment of the black figures, which reflected the long history of black treatment on American stages. Nevertheless, Root's story bravely and innovatively moved forward to flesh out its figures, presenting them as relatively three-dimensional, workaday people. In a small way his approach harked back to Harrigan's understanding portraits. Operettas such as *Deep River* and *Show Boat* had attempted to present blacks realistically during the twenties, as had some later revues. But *Cabin in the Sky* was virtually the first musical comedy to join the list.

The original production featured an incomparable cast, with the biggest black star of the era, Ethel Waters, as Petunia, Dooley Wilson as Joe, Todd Duncan (the original musical Porgy) as The Lawd's General, and Rex Ingram as Lucifer, Jr. No less than Katherine Dunham herself assumed the role of the hussy, Georgia Brown. Miss Waters's magnetic presence undoubtedly helped the show to overcome some prejudices and run twenty weeks.

The musical was revived joyously on the summer circuit in 1953 and off-Broadway in 1964. Curiously, the *Times* assailed this last revival. The paper and Atkinson had long ago taken the lead in denouncing operetta as reactionary and old hat. Now, with Atkinson retired, the paper and its drama critic, Howard Taubman, pandered to professional agitators by attacking the show. Taubman wailed, "A fantasy in which the Negro is treated like a simple child of nature, moving and talking and sinning and shouting in ways that have become annoying stereotypes, is not so palatable as it was in the seemingly more innocent year of 1940 when this musical was new." That many simple children of nature—black or white—still exist and still move, talk, sin, and shout in similar ways was conveniently forgotten. So were the facts that 1940 was not innocent and that Atkinson, whatever his blind spots, was a sensitive and perceptive critic.

Even Atkinson himself was not to forget that three months after he wrote about *Cabin in the Sky* he concluded his review of *Pal Joey* by asking, "Can you draw sweet water from a foul well?" He labeled the story "odious" and insisted the "evening offers everything but a good time." Enough theatregoers disagreed to allow *Pal Joey* to play for 374 performances. They were pleased with the very story that offended Atkinson. Nothing quite like it had ever been attempted before in musical comedy.

Unsavory characters occasionally had been granted center stage in earlier shows—in *The New Yorkers*, for example—but their ugly side had been dressed up in garish colors and furnished with all manner of comic accessories. John O'Hara, who molded the libretto from raw materials in his *New Yorker* short stories, would have none of this. He stripped his characters bare and presented them to Broadway in a cynical, hard-hitting, yet not wholly unsentimental saga of bought love. His heroine was hardly the demure young innocent that had long been musical comedy's stock-in-trade. She was a steely, selfish, past-her-prime matron who would pay for the attentions, if not the affections, of a seedy, empty, equally selfish young dancer. Although at one point she proclaims that she was "bewitched, bothered and bewildered," her eyes are really wide open. So are his. Still, Vera can be generous to her Joey, not merely setting him up in a lavish apartment of his own but writing the checks that allow him to set up his own nightclub. Her largesse doesn't stop Joey from playing around with younger, more attractive girls, but Vera is willing to look the other way, up to a point. Only when she is threatened with a sordid blackmail attempt does she decide enough is enough.

Among those in sharp disagreement with Atkinson was the often cantankerous and feisty critic for the *New Yorker*, Wollcott Gibbs. His response to *Pal Joey* was handclapping delight, which stemmed in large measure from the show's characters—"living, three-dimensional figures, talking and behaving like human beings." O'Hara had not merely created interesting, believable people but he had given them some of the best dialogue Broadway musical comedy had ever heard. Though their conversation was finely honed for theatrical purposes, it was nonetheless taut, sharp, and accurate. When Vera finally gives Joey the brush-off, she attempts to be polite. But Joey will have none of it. He turns nasty, and he and Vera go at each other with verbal daggers.

> *Vera.* I have a temper, Beauty, and I want to say a few things before I lose it.
>
> *Joey.* Lose it. It's all you've got left to lose.

Audiences also enjoyed the songs, a joy in this case that even Atkinson shared. They represented Rodgers and Hart at their best. "Bewitched" and "I Could Write A Book" have proved durable favorites, but almost every number in the show attested to Rodgers's melodic genius and Hart's unique way with words. Indeed, the very nature of the show gave Hart a field day to put in rhyme his amused, misogynistic view of the world. A few of the nightclub numbers may have been discreet bows to musical-comedy necessities, but even in those Hart generally found ways of making his lyrics pertain to the story.

There were, of course, other reasons for *Pal Joey's* popularity, although they would have mattered little had the book and songs not been so strong. The musical was given a sumptuous production and was impeccably cast. Vivienne Segal, a fine singer, was able to show off her superb comic gifts whenever O'Hara's or Hart's black humor allowed her. And a rising dancer, Gene Kelly, was rocketed to stardom as Joey.

The last of this trio of superior musical comedies, *Lady in the Dark*, was also 1940–41's last musical. Like *Cabin in the Sky* and *Pal*

Joey before it, the new show offered one of Broadway's biggest stars as a major inducement. For *Lady in the Dark,* that star was the radiant Gertrude Lawrence, certainly one of the most beloved of all Broadway performers. Yet the show was not conceived with her in mind. Moss Hart had undergone psychoanalysis and had decided to write a play that drew on his experience. Initially at least, he envisioned it as a vehicle for Katharine Cornell. But, as noted earlier, he grew to realize his story cried out for music. Although he changed tack in mid-work, his original conception gave his libretto a solidity, brilliance, and sophistication it might otherwise have missed. Indeed, when *Lady in the Dark* finally reached Broadway, Hart's contribution was so impressive that Atkinson urged his readers to take note. He insisted the entertainment was a musical play, not a musical comedy, adding, "What that means . . . is a drama in which the music and splendors of the production rise spontaneously out of the heart of the drama, evoking rather than embellishing the main theme." Atkinson's observation was pertinent but also, in a way, meaningless. Time and again, albeit more often with operetta than with musical comedy, the claim had been put forth that song and story were so well integrated that a new appellation was required or that a whole new genre had been created. The cry was to be forthcoming again and offered with renewed vigor three years later when *Oklahoma!* premiered. But all book shows are, after all, essentially musical plays. Their qualities, including their ability to integrate all elements skillfully, help to determine whether they are good or indifferent musical plays, not whether they are operetta or musical comedy.

Lady in the Dark was a superb musical comedy, with, perhaps, one minor flaw. Hart's libretto centered on Liza Elliott, the elegant but icy editor of a celebrated fashion magazine. For all her public aplomb, she is a privately unhappy woman. She is, in fact, unhappy enough to submit to analysis, in which she tells the doctor of her dreams. These dreams disturb her because, while they are filled with familiar people, those people behave in unfamiliar ways. Most of the dreams involve the four men in her life: her unsatisfactorily married lover, who has pushed her to the top; a glamorous but vacuous Hollywood dreamboat; the magazine's faggoty photographer; and her sassy, strong-willed advertising manager. With her analyst's gentle prodding, Liza delves into her past to learn why she acts the way she does. One outcome of all her probing is her recognition that the man she truly loves is her advertising manager.

Hart's dialogue eschewed O'Hara's knife-edge cruelty and toughness, which would have been wrong for the play. It was as elegant as Liza, while it was just as knowing and credible. At times Liza speaks whole paragraphs without becoming wordy. For example, after assuring her analyst that she has only contempt for ladies who relieve their boredom and frustration by paying him "so much per hour," she continues, "Let me get one thing straight, Dr. Brooks. There's nothing strange in my life. I have no queer twists. I am doing the kind of work I care most for and I am enormously successful at it. My love life is completely normal, happy and satisfactory. I wish there were some little phobia for you to gnaw at. But there isn't."

Ira Gershwin's lyrics were Gershwin at his best—meditative when situations required them to be, witty when circumstances called for humor. His lines were literate but sprinkled with his identifiable shorthand, as when Liza proclaims "gloom can jump in the riv'!" His lyric for "The Saga Of Jenny," in which, during one of the dream sequences, Liza recounts the misadventures of a girl who thought she knew her mind, remains as fresh and hilarious after repeated hearings as it was at first. And for young Danny Kaye, Gershwin provided a tongue-twisting list of Russian composers in "Tschaikowsky."

Gershwin's words were set to Kurt Weill's evocative, appropriate, and eminently listenable music, if not music that stuck in some corner of the mind and demanded to be whistled. None of *Lady in the Dark*'s songs has found a permanent niche in the public's affection. Kaye still employs "Tschaikowsky" as a tour de force in personal appearances, "The Saga Of Jenny" is still performed as a comic number, but only "My Ship," the theme played throughout the evening and the only song in the show not sung in a dream number, is occasionally heard— and that on rare occasions. This may well explain why *Lady in the Dark*, unlike *Cabin in the Sky* and, since its ragingly successful resurrection in 1952, *Pal Joey*, has enjoyed no major revival. Playgoers cannot connect it with any "standard."

Lady in the Dark's mounting was the most lavish of the season. Harry Horner's colorful sets were turned, as the audience watched, on four revolving stages. Irene Sharaff and Hattie Carnegie designed stylish, tasteful, and sometimes (for dream sequences) outlandish costumes.

Lady in the Dark shared with *Cabin in the Sky* and *Pal Joey* one other innovative feature. The dancing was increasingly balletic. In fact, *Cabin in the Sky* went so far as to introduce folk dance motifs into its choreography. All three shows retained some standard routines, precision numbers or tap dances, but they were clearly moving away from these old standbys. Indeed, in assuming the overall direction as well as most of the choreographic chores for *Cabin in the Sky*, George Balanchine was pointing the way even farther into the future.

In another move away from musical-comedy tradition, all three stories centered on older-than-ordinary principals. Pink-cheeked, fresh-out-of-school lovers were shunted aside. Although perhaps only Vera and Liza could be perceived as middle-aged, and then barely so, they were appreciably beyond the wholesome, wide-eyed innocents so long musical-comedy standbys.

Lady in the Dark's glittering star, eye-filling production, and intelligent words and music combined to earn it a run of 467 performances. It barely missed that golden five hundred mark that the less deserving *Panama Hattie* surpassed. Whether Broadway took that as a lesson or whether the war, which broke out at the end of the year, determined the nature of shows, the 1941–42 season brought nothing comparable to the three innovative musicals of 1940–41. *Best Foot Forward, Let's Face It, Banjo Eyes,* and *By Jupiter* had commendable aspects, but they were cast in more or less traditional molds, with nothing about them that could be hailed as adventurous or of exceptional quality. The following season, 1942–43, had still fewer good musical comedies—only *Something for the Boys* was a hit. But toward the close of the season, on March 31, 1943, *Oklahoma!* arrived and changed Broadway's notions of musical entertainment.

12

Musical Comedy
in the Heyday of
the Musical Play

If *H.M.S. Pinafore* is considered the most influential musical in our theatre's history, it achieved that position by opening almost all our stages to lyric entertainments, by prompting a steady outpouring of musicals, and, of course, by providing a model for contemporary librettists and song writers. But it cannot be said to have radically changed the nature of musical-theatre writing, in the sense that there had been little before it that could be changed. *Pinafore* established the nature of our musical theatre more than it altered it. The show that most notably changed America's thinking about the nature of musical theatre was almost certainly Richard Rodgers and Oscar Hammerstein II's *Oklahoma!*

In the widespread, exhilarated ballyhoo that followed *Oklahoma!*'s tremendous success it was generally proclaimed that the show had initiated a new lyric genre, the musical play, and that musical plays were characterized by an unprecedented integration of song and story. Forty years of hindsight has suggested that such pronouncements, however honestly made, were misguided and exaggerated. As we saw earlier, neither the term "musical play" nor the integration of song and story was all that new. Cogent arguments could be mustered to demonstrate that any number of shows—mostly operettas—had skillfully integrated song and story from Gilbert and Sullivan on. At best, *Oklahoma!* could lay legitimate claim to have carefully woven a new element—dance—into the artful fabric of the modern musical. No longer would singers sing and then go into their dance, a purely decorative dance at that. Agnes de Mille's ballet in *Oklahoma!* brought to life the heroine's dream and provided her motive for refusing the hero's invitation to a box social. It was part of the story.

The idea that integration was something spanking new and desperately needed took hold of Broadway's thinking. Integration of songs, story, and dance became more than a catchphrase; it became the keynote of most musicals that followed *Oklahoma!* into New York. In fact, it became so fashionable to integrate all these elements, that ballet was sometimes injected when it served no dramatic purpose, and sometimes even when it hindered a proper unfolding of the story. Broadway's reasoning seemed to be that integration required ballet. It was, in a contemporary sense, "bussed in" whether it helped or not.

But while *Oklahoma!*'s integration of song, story, and dance was noisily celebrated on all sides, the artists associated with the show had quietly achieved additional integration that received far less attention, an integration of tone and style that was just as important as the textual integration. This too was not new, although it was probably rarer in older shows. Think, for example, of the interpolated songs of bygone times that might have related to the character performing them but nevertheless were composed by an author with distinct musical mannerisms.

With time, it became apparent that *Oklahoma!*'s real importance lay elsewhere. To begin with, the show made the American musical theatre look at America's own heritage for inspiration. Nostalgic memories of the American past hereafter provided a fertile field for librettists. *Show Boat*, of course, was given a deep bow in this respect, as well as for its integration of song and story. Articles extolling the glories of the new school of musical play regularly paid tribute to Kern and Hammerstein's masterpiece. And there had been other, less readily remembered examples. But with *Oklahoma!* Americana became the rage of our musical theatre.

More importantly, as Richard Rodgers himself wrote so perceptively, "the chief influence of *Oklahoma!* was simply to serve notice that when writers came up with something different and if it had merit, there would be a large and receptive audience waiting for it." Untouchable subjects became acceptable, and the treatment of these subjects, even when they were painful, could be refreshingly honest. Much was made at the time of the hero's killing the villain on stage in *Oklahoma!* This too was not new. Porgy had killed Crown in *Porgy and Bess.* *The Vagabond King* had even had an on-stage suicide. But while the claim to originality was once again exaggerated, *Oklahoma!* by virtue of its huge popularity—a popularity in no way diminished by an unpleasant scene—did open doors.

In one respect, this willingness to face unpleasantness had its unfortunate side. Sometimes whole stories became somber and ugly. Honesty of treatment soon led to vulgar language, which in careless hands was used merely for its shock appeal. All of this became evident more in modern operetta (which is what "musical plays" are) than in musical comedy. For the most part, musical comedy, because it has always set out to have fun, had little to do with this darker side of things. True, it more often than not lost the lighthearted lunacy that had characterized it, and there were many who rued the loss of this "dated dementia," as the Philadelphia *Inquirer's* William B. Collins has described it. On the whole, however, the best musical comedies to premiere in the years after 1943 profited from the new school's integrity without getting bogged down by its flaws.

Of course, not all musical comedies subscribed to the new school. Many an old-fashioned offering, with its line of pretty girls, its great comics, its snappy dances came along and proved ragingly popular. Witness *Early to Bed, Mexican Hayride, Follow the Girls, Are You with It?, Gentlemen Prefer Blondes, Call Me Madam, Wish You Were Here, Silk Stockings, Ankles Away, Wildcat, Do Re Mi, Little Me, Mame*—the list goes on and on. If a few of these were commercial failures despite long runs, so were many of the better operettas. Theatrical economics turned into a horror story during these years.

Mexican Hayride typified this class of entertainment. One of the luckier shows, it realized a handsome profit on its long run, even though it was the most expensively mounted musical since Ziegfeld's heyday. Herbert and Dorothy Fields provided a book that was pure hokum—something about a clownishly resourceful fugitive who has an unwelcome spotlight cast on him when an American lady bullfighter throws him her bull's ear. What drew packed audiences were Mike Todd's gaudy production, with its brigade of beautiful, often scantily clad girls, Cole Porter's often jaunty, always hummable songs (most memorably "I Love You"), and the sidesplitting antics of Bobby Clark. What made Clark's clowning so great is difficult to describe. On paper his gimmicks and routines seem much like the stock-in-trade of many lesser comedians. Nevertheless, his painted-on glasses, his ever-present cigar, his stubby cane, his outrageously rolled rs and caterwauling elicited instant smiles of recognition and expectation from audiences. Highpoints of his clowning in *Mexican Hayride* were the disguises he employed as a fugitive. In one scene he became part of a native Mariachi band. Having been told Mariachis were Mexican minstrels and thinking Mexico far behind the times, he joins the band costumed as a medieval troubadour, playing classical music on a flute in counterpoint to the Mexicans' Latin rhythms. Later he pretends to be a Mexican Indian squaw, sporting not merely a dress and a wig, but a set of false buck teeth along with his cosmetic glasses and his cigar. When he turned his back to the audience he disclosed a papoose, who also had pointed-on glasses and a cigar. Clark scarcely needed disguises for laughs. Another comic moment in the show came when he put on a minature pole-vaulting exhibition, using his sawed-off cane. The era of great clowns was drawing to an end, and with their disappearance musical comedy would lose a certain wondrous zaniness. But the forties and fifties saw the last great survivors still in peak form. A few younger comic talents appeared—Nancy Walker, Phil Silvers, Sid Caesar—but they soon found television more lucrative.

Another group of musical comedies was essentially just as old-fashioned but was touched by the new school's influence. This influence was especially noticeable in the shows' dancing. *Look Ma, I'm Dancin'*, *High Button Shoes*, and *Can-Can* all offered trite, silly stories and superb dancing. Jerome Robbins's "Mack Sennett Ballet" for *High Button Shoes* remains the greatest bit of comic choreography our musical stage has yet produced. Robbins was able to incorporate this lunatic spoof of silent film comedy into the show because *High Button Shoes* followed the musical play's fashion of appropriating turn-of-the-century America for its setting. A new breadth of subject matter was also evident in these shows, and with it a new honesty of treatment. *Plain and Fancy*, for example, took a thoughtful look at Pennsylvania's Amish that was as compassionate as it was funny.

Sometimes the debt was less obvious or tangible. Often it seemed to be little more than a pervasive style of mounting, a style fertile in wit and imagination. This was usually the brainchild of the director, but the best directors somehow managed to impart their spark to designers and casts as well. *Hello, Dolly!*, one of the most successful of all American musical comedies, is a choice example, although it is from a later period. It owed no small measure of its popularity to its title song, a song that for months Americans could not seem to hear enough of. And then there was Carol Channing's captivating performance as Carol Channing masquerading as Dolly. But the real glory of the original production was Gower Champion's inventive stylization with its mock hauteur and tongue-in-cheek period nostalgia. Stripped of Champion's touch, *Dolly* was really an ordinary show. Its sole inherent strengths were those borrowed from its source, Thornton Wilder's *The Matchmaker*.

A few shows defy easy classification. Several fundamentally traditional musical comedies put it all together in a way that cried out "new." *Where's Charley?*, a musical version of *Charley's Aunt*, which featured Frank Loesser's offbeat songs, including the still popular "Once In Love With Amy" and Ray Bolger's inimitable vaudeville stepping, was a largely satisfying blend of yesterday and today. Its director and librettist, George Abbott, also was director and librettist for *Damn Yankees*, which had fun with the devil and baseball. *Damn Yankees* offered the sort of book that would have raised no eyebrows in the thirties or forties and that was coupled with Richard Adler and Jerry Ross's insistently contemporary yet invitingly melodic songs such as "Whatever Lola Wants," "Heart," and "Two Lost Souls" and Bob Fosse's fresh, high-voltage choreography, especially as interpreted by Gwen Verdon.

Probably the best of these somewhat unclassifiable musical comedies was another of our theatre's most memorable triumphs, *Annie Get Your Gun*, produced by Rodgers and Hammerstein. A theatricalized biography of the famed sharpshooter Annie Oakley, and her real-life romance with Frank Butler, its Herbert and Dorothy Fields book was a standard laugh- and tears-filled girl meets, loses, and retrieves boy yarn. Its Irving Berlin songs were the sort of direct, catchy ones Berlin had been composing for decades. A friend of the composer's complained to him that *Annie Get Your Gun* was old-fashioned. Berlin is said to have retorted, "Yes, a good old-fashioned smash." Yet one reason for its runaway success was the perception that it was something of a new school musical. On a most superficial level, its recourse to a turn-of-the-century setting aligned it with the new operetta's penchant for nostalgic American backgrounds. Berlin packed his score with showstoppers; in the traditional meaning of the term, but in the newer way of things almost all the songs had logical reasons for being where they were and actually moved the story along. Both the Fieldses' book and some of Berlin's songs probed the characters' feelings with more understanding and depth than had been common before. And their probing and the songs' integration seemed to come about with an almost careless ease, without self-conscious striving. Of course, Berlin's great songs were so good that they have enjoyed a long life away from the show. "Doin' What Comes Natur'lly," "The Girl That I Marry," "They Say It's Wonderful," "I Got The Sun In The Morning," and "There's No Business Like Show Business" have all become standards. No small part of their original success was attributable to Ethel Merman's delivery, but Mary Martin endeared them to theatregoers in her own way when she headed the road company.

For the most part, however, the greatest musical comedies of the late forties, the fifties, and the very early sixties seemed determinedly progressive, determined to advance, or at least change, the very nature of musical comedy. Their cry became "sing me a song of artistic significance." The best shows were indeed textually and stylistically integrated, their dialogue and lyrics fresh and colloquial, the development of their stories honest and of their characters realistically three-dimensional. A few truly inventive works failed for one reason or another—usually because of book problems. *The Day Before Spring* (with its all too neglected Frederick Loewe music) and *Out of This World* (with its equally neglected Cole Porter score) were beset by heavy-footed books. *Flahooley,* wonderfully witty and cherishingly melodic, was defeated by an overwrought story. *Love Life* attempted a unique amalgam of book musical, vaudeville, and fantasy—and tripped over itself. Oddly enough, when Rodgers and Hammerstein ventured away from musical play or operetta into the realm of musical comedy, their special touch eluded them. *Pipe Dream* and *Me and Juliet* failed artistically and commercially; *Flower Drum Song* succeeded to no small extent because of Rodger's score. After Hammerstein's death, Rodgers briefly became his own lyricist when he added words to his splendid score for *No Strings,* another musical beset by an inferior libretto and a theme (a romance between a white man and a black girl) that touched a raw nerve with some playgoers.

Happily, these failures were few in number—a number far surpassed by the masterful musical comedies that proved both artistic and commercial triumphs. *One Touch of Venus, On the Town, Finian's Rainbow, Guys and Dolls, Kiss Me, Kate, Gypsy, How To Succeed in Business Without Really Trying* all remain enduring joys. Since *One Touch of Venus* may be the least revived and therefore the least familiar of these shows, let's take a longer look at it.

With Mary Martin, Kenny Baker, and John Boles as its stars, and a fine comedienne, Paula Laurence, in support, *One Touch of Venus* opened in New York in early October 1943. As such, it was the first important new wave musical comedy. *Oklahoma!* had premiered less than seven months before. Yet there was nothing tentative about *Venus,* none of the gawky missteps so common in pioneering works. It boasted an accomplished Kurt Weill score, pyrotechnically brilliant Ogden Nash lyrics, and a wise, mocking S. J. Perelman-Nash libretto, all cleverly meshed to satisfy the growing insistence on integration. And part of the show's story was moved forward by two of the newly requisite Agnes de Mille ballets. Indeed, the dancing was so demanding that it required a fine young ballerina, Sono Osato, to assume Venus's role in the ballets.

The story was hardly new, merely the latest retelling of the Pygmalion-Galatea legend. Broadway had embraced another version,

Adonis, nearly sixty years before. (It is a bit unnerving to realize that *One Touch of Venus* itself is now approaching its fortieth birthday.) But whereas *Adonis* had depended on burlesque absurdities and the visual buffoonery of a capital clown for its laughs, *One Touch of Venus* told its story in a straightforward, down-to-earth manner, drawing its humor from the very human foibles of its characters and the sometimes sly, sometimes patently barbed wit of its lines.

One Touch of Venus begins in a swank modern art gallery. The gallery is the personal toy of Whitelaw Savory, an extravagantly rich eccentric who delights in teaching his unorthodox ideas to his students. New art, he tells them, is the only "true art," adding, "The old masters slew art." Thus the musical's very first lines can be interpreted as an indirect plug for the new school of musical theatre and a dig at almost everything that went before. Of course, like the musical he walks through, Savory is not totally consistent in his art preferences. Some vestiges of the past remain to claim his attention. He announces that he at last has secured a classic Greek statue, one that he has had his eye on ever since he first read about it. And he candidly confesses that more than a smidgin of his interest in the marble derives from its resemblance to a girl he once loved and lost. His students can't see the connection between the classic statue and modern art, but Savory's loyal, wisecracking secretary tells them they will have to consider Venus of Anatolia an exception to any number of rules. She notes that Venus was a goddess in a world dominated by gods but had only to open up her bodice to equalize the odds. (A little woman's lib decades before the term was invented, although not quite the way modern women would have it.) The statue is unveiled with all due ceremony, and the crowd disperses.

A young man enters. He is Rodney Hatch, a barber come to shave Mr. Savory. Rodney is about to be engaged, so when Savory insists that his Venus is the most beautiful woman ever conceived, Rodney disagrees. Savory is called away to the phone. To prove his point, Rodney decides to put the engagement ring on the statue's finger. He is certain his Gloria's fingers have more grace and delicacy than Venus's. He puts the ring on one outstretched finger and in a blinding flash of light the statue comes to life. Rodney demands his ring back, but Venus, fearing its return will mean her own return to stone, refuses. Befuddled and terrified, Rodney runs off. So does Venus. When Savory reappears and finds both the barber and the statue gone, he sounds an alarm.

Back in his apartment Rodney is on the phone with his fiancée, Gloria Kramer, trying to explain that a statue stole her engagement ring. For some reason, Gloria finds this hard to believe. When she hangs up, Rodney then tries to make her picture understand how much he loves her. But his comparisons are odd and a field day for Freudians, to say the least—for example, "I love you more than a wasp can sting," or "more than a chilblain chills." His song is no sooner finished than he is startled to discover Venus in his room. Her arrival coincides with Gloria Kramer's calling back. It is all Hatch can do to keep Venus from talking to Gloria. But his troubles have just begun. His landlady appears, indignant at his having a strange woman in his room. A wave of Venus's finger disposes of the troublesome old lady. Once again, Rodney runs off.

Alone in Radio City, Venus confesses "I'm A Stranger Here Myself" and that she has much to learn about twentieth-century ways. She then, in the first ballet, proceeds to bring window mannequins to life and allows them a fling with some passing sailors. To get out of her classic robes, she also steals an ensemble that a window dresser has left behind. Her actions cause mayhem and bring the police. Luckily, Savory happens to pass by. Discovering his Venus is well and surprisingly alive, he immediately begins to court her, hinting at his long lost love. Venus will have none of this starry-eyed sentimentality. Love to her is "the triumphant twang of a bedspring." When Savory all too eagerly misconstrues her remark she hurriedly puts him in his place by assuring him she was proffering an opinion, not an invitation. She runs off to find Hatch while Savory mooningly asks if the "West Wind" can find his old flame.

Hatch has come to the Mid-City Bus Terminal to meet Gloria and her mother. He and Gloria celebrate the joys of traveling "Way Out West In New Jersey." Their happy plans are thrown for a loss by Venus's arrival. She soon sets the lovers feuding. Gloria issues an ultimatum: get her ring back within twenty-four hours or the wedding is off. Rodney starts to berate her, but when Venus agrees with him, he stalks off, bellowing that she had no right to talk about his girl that way. Baffled by Hatch's obtuseness, Venus seeks Savory's advice. But in admitting to having a "Foolish Heart," she makes Savory realize that Hatch is his unintentional rival. Savory and his hirelings determine to get rid of Rodney. They congregate at his barbershop where they soon find themselves examining the difficulties women have made for them. "The Trouble With Women," they conclude, is men. The philosophizing done, the hirelings get down to business, locking Hatch in the cellar while they look for the statue. Gloria arrives and is promptly tied up in the barber chair.

When the men leave, their search fruitless, Venus appears. Gloria is not happy to see her. An argument ensues, and Venus, with another wave of the finger, causes Gloria to vanish. Gloria leaves behind a compact, which Venus appropriates. Releasing Hatch from the cellar, Venus soothingly urges him to "Speak Low." For the first time Hatch is rewarded with a glimmer of awareness that he loves the ex-goddess. Savory has invited everyone to a party, hoping someone will let slip the statue's whereabouts. Mrs. Kramer spots Venus with Gloria's compact. Since no one can say what has happened to Gloria, Venus and Hatch are carted off to jail for having kidnapped her.

The second act opens in Savory's luxurious bedroom. His luxury fails to ensure his privacy, however, for the room is invaded by an Anatolian crackpot who is bent on returning the statue to his homeland. Savory sends him after Hatch, then complains to his secretary about his hard life. The secretary, Molly, can't agree. Life with thirty-five million in the bank isn't all that difficult. Besides, "You can huddle with your memoirs, and boy! what memoirs them was!" when you are "Very Very, Very" rich. Meanwhile, things are unpleasant enough in jail for Venus and Rodney, except, of course, that they are together. However, when the demented Anatolian begins to menace them, it proves too much. Venus flings open the doors, and the pair escape. They take refuge in a lavish hotel suite, only to have Gloria barge in demanding her ring. Seeing Rodney is with Venus, she storms out in a fury. "*Sic Transit* Gloria Kramer," Venus sighs. Acknowledging they are in love, Rodney promises Venus their romance will flower even further. He paints a picture of how idyllic domesticity will be five years hence, when they celebrate their "Wooden Wedding." His vision turns into flesh and blood, and in the ballet, "Venus In Ozone Heights," suburbia's humdrum daily routine is brought home to the goddess.

Back at Savory's gallery, Savory, Molly, the mad Anatolian, and Rodney stand before Venus's empty pedestal. The Anatolian is ready to kill Savory, but suddenly in another blinding flash the statue is restored to its stand. As the rest of the group go their separate ways, a saddened Rodney stands in front of the statue. Suddenly, a young girl appears. Though she is dressed in country clothes and is a bit awkward, she is the image of the goddess. Hatch rushes forward to introduce himself. She is about to tell him her name when he stops her— "You don't have to tell me. I know."

If *One Touch of Venus* is to be faulted, that fault may well lie in Kurt Weill's inability to provide an outstanding song or two. At performances, Mary Martin, dressed in a lacy Mainbocher negligee and leaning against the back of a chair, serenaded her audiences with a bemused soliloquy, "That's Him!" and won nightly encores. "Speak Low," a beguine, has remained marginally popular. But none of the musical's songs has ever commanded the forefront of popularity. This failure calls to mind the same shortcoming in *Lady in the Dark* and thus possibly explains both musicals' otherwise inexplicable neglect on revival stages.

No such neglect has deprived happy playgoers of most other masterpieces. Although some have been revived more regularly than others, hardly a season passes without notable revivals of a sizable majority of these shows. They have been standbys of summer theatre, scholastic playhouses, and those hybrid abominations, dinner theatres. *On the Town* and *Finian's Rainbow*, though both have been remounted for Broadway, are probably the most revival-resistant: *On the Town* because of its demanding choreographic requirements and *Finian's Rainbow* because so much of its original topicality has become dated. By coincidence, they were the first gems to appear after *One*

Touch of Venus. On the Town got in under the wire at the very end of 1944; *Finian's Rainbow* premiered just over two years later in early January 1947.

Leonard Bernstein and Jerome Robbins's ballet, *Fancy Free*, provided the seed for *On the Town*. Both were rhythmically alive, capturing the pulsating, sometimes frenetic, drive of the war effort. In the show, three sailors on leave in New York from World War II seek a girl whose picture caught their eye in a subway ad. The musical had a hilarious book and lyrics by Betty Comden and Adolph Green and riotous performances by Comden, Green, and Nancy Walker. The original ballet was fleshed out attractively, but dancing continued to be the show's backbone. Among Bernstein's melodies, "Lucky To Be Me" and "New York, New York" are still heard and recognized. Nine years later, Bernstein, Comden, and Green (along with Joseph Fields, Jerome Chodorov, and George Abbott) created a lighthearted romp for Rosalind Russell, *Wonderful Town*, a musical version of *My Sister Eileen*.

Finian's Rainbow was something of an oddity, a propaganda piece as well as an entertainment. The propaganda was blatantly left-wing (has anyone ever heard of a right-wing musical on Broadway?), but it was so sunny, so warm-hearted, and so witty that only the most archly conservative curmudgeon could object. At its best it was an imaginative and simple plea for racial understanding. A daffy Irishman, his patient, loyal daughter, a whimsical leprechaun, and an entrancing, dancing deaf-mute help to teach Senator Billboard Rawkins and his Missitucky cronies the price of bigotry. Time has dulled the edge of many of the musical's barbs, especially for younger playgoers unfamiliar, say, with the real-life antics of Mississippi's Senator Bilbo and Representative John Rankin. But the luster of Burton Lane's richly melodic score remains undimmed. The tender, evocative "How Are Things In Glocca Morra?" quickly became one of the hits of the season. Even with repeated hearings Lane's whole score retains a unique freshness, and songs such as "If This Isn't Love," "Look To The Rainbow," and "Old Devil Moon" still pop up at intervals. In the original, Ella Logan (despite her Scottish accent), Albert Sharpe, Donald Richards, and, most of all, David Wayne as the leprechaun made the denizens of the imaginary Missitucky real and unforgettable.

Kiss Me, Kate arrived almost two years after *Finian's Rainbow*, and another two years elapsed before *Guys and Dolls* raised its first Broadway curtain. *Kiss Me, Kate* is generally regarded as Cole Porter's finest accomplishment; and it was. Often overlooked is the fact that the musical was also Sam and Bella Spewack's finest achievement. Their libretto centered on a divorced acting team who find themselves cast as Kate and Petruchio in a revival of *The Taming of the Shrew*. The musical begins on the day the two performers are celebrating an anniversary—the first anniversary of their divorce—and then alternates between their backstage bickering and their sometimes unprofessional onstage rows. By the final curtain, when Kate kneels humbly before Petruchio, the actor has tamed and reclaimed his ex-spouse. A subplot, carefully interwoven with the principal story, told of the rocky romance of two other performers, a romance jeopardized by the young man's gambling.

Because the two stories were developed logically and coherently and because their humor derived entirely from the situations and characters, the Spewacks' work was acclaimed as a shining exemplar of the new school of lyric theatre. Porter's music and lyrics moved deftly and gracefully between the show's two worlds, not merely decorating the story, but commenting on it and moving it along. For once, Porter's songs seemed to belong to the show. The composer developed his lyrics from Shakespeare's lines and, most effectively and affectingly, let Shakespeare be his lyricist in the touching finale, "I Am Ashamed That Women Are So Simple." Many of the show's songs—the haunting beguine, "So In Love"; the satiric waltz, "Wunderbar"; the thumping anthem, "Another Op'nin', Another Show"—have enjoyed lively, long-lasting, and independent vogues. *Kiss Me, Kate*'s textual felicities were made all the happier by a rousing production that included Lemuel Ayres's stylized sets, Hanya Holm's leaping choreography, and, perhaps most of all, a perfect cast led by the greatest of contemporary performers, Alfred Drake. Drake would have been a rarity in any age. He was a darkly handsome performer who combined magnificent singing and intense acting with a singular gift for high comedy.

Comedy was the high point of *Guys and Dolls*, a tough-talking, big-hearted musicalization of Damon Runyon's sagas of New York low life, notably "The Idyll of Miss Sarah Brown." From these tales Abe Burrows selected two guys, both professional gamblers, and two dolls, a singer in a sleazy nightclub and an idealistic Salvation Army belle. Their romances and adventures in Manhattan's netherworld provided his plot. As a result, Burrows's humor was leathery and sometimes brazen. When one of the gamblers, having just described a pestiferous detective as "that lousy Brannigan," realizes the detective has over-heard, he turns to the cop and assures him he did not have the detective in mind and that "There are other lousy Brannigans." So adroit was Burrows's writing that he effectively created a unique, private world, as his subtitle, "A Musical Fable of Broadway," hints. That world was peopled with characters sporting such distinctive monickers as Sky Masterson, Nathan Detroit, Nicely-Nicely Johnson, Harry the Horse, and Arvide Abernathy.

Frank Loesser's lyrics and music matched Burrows touch for touch. No modern Broadway composer surpassed Loesser's knack for inject-ing as much wit into his melodies as into his words or of capturing the rhythms and tensions of everyday drama in his musical phrases. Several of his best numbers—"Adelaide's Lament," "The Oldest Es-tablished"—are masterful comic vignettes, concisely etching charac-ters and milieus. Loesser could also be effectively sentimental, softly so in "More I Cannot Wish You," more fibrously in "My Time Of Day" and "I've Never Been In Love Before." "If I Were A Bell" and "A Bushel And A Peck" were offbeat love songs infused with Loesser's witty slant on Broadway romance. As director, George S. Kaufman stamped his own brand of misanthropic humor on the production, elic-iting marvels from a stellar cast headed by Sam Levene, Robert Alda, Vivian Blaine, and Isabel Bigley. Michael Kidd's vivid, picaresque choreography wove itself into a seamless whole.

All these musicals were totally contemporary in setting, and, ex-cept for *Finian's Rainbow*, all did or could have taken place in New York. Technically, *Kiss Me, Kate* unfolded during a tryout in Baltimore, but a few nasty digs at that city aside, it could have easily been set on Broadway. While many lesser musical comedies were responding to the new vogue for older American backgrounds, these masterworks ignored the fashion. They cavorted across musical comedy's traditional terrain, though not as topically and brashly as in the thirties. Only in the late fifties did several superior pieces fall in line.

Both *The Music Man* and *Fiorello!* spilled over at times into operetta, perhaps because they returned to bygone worlds for their inspiration. Indeed, *The Music Man*, moving farther into the past than *Fiorello!*, also moved much closer to operetta. But its roseate glimpse of turn-of-the-century Iowa was not truly the high-flown, arch romanticism that was operetta's stock-in-trade. Meredith Willson, who wrote the entire show—book and songs—built his piece around a lovable con man, a honey-tongued traveler capable of fast-talking a small farm community into buying a band load of musical instruments for its schoolchildren. The town's young librarian sees through him and ultimately brings about his reformation. She wins his hand in the process. Willson's book painted a charmingly deceptive picture of imagined innocence and harmless guile. It was as wholesome and lovable as grandma's apple pie.

So were the songs, although they were far more inventive. Willson returned to all the proper period forms—thumping marches, gentle lullabies, closely harmonized barbershop quartets, lively buck and wings, lazy soft shoes—with wit and imagination. The musical's most popular song, the hero's "Seventy-six Trombones," was an exuberant march. Later in the evening Willson reemployed the same melody in the heroine's gentle "Goodnight, My Someone," thus quietly underscoring the principals' relationship. "Trouble" was an ingeniously contrived modern-day patter song. More ingenious still was the opening number, "Rock Island," in which a group of itinerant salesmen compare notes. The heavily accented rhythm of the unrhymed lyric caught the clickety-clack of the train the men were riding and thus set the mood. Robert Preston, previously better known for his work in Hollywood westerns, was an irresistible bunco artist, and the lovely voiced Barbara Cook was his enchanting vis-à-vis.

Something of a similar rose-colored gel threw its glow on *Fiorello!*

After all, Fiorello was obviously Fiorello La Guardia, the pugnacious, bantam-sized mayor who tried to sweep New York politics clean. Both Jerome Weidman and George Abbott's book and Jerry Bock and Sheldon Harnick's songs naturally cast La Guardia and his friend in the brightest, purest light. But the mayor's crooked political enemies came out of the show as likable grafters. It was, the authors seemed to be saying, all good fun. Certainly it was the most joyous fun for audiences. Bock and Harnick added immeasurably to everyone's pleasure with their superlative songs, which ranged from a beguiling period waltz, "Till Tomorrow," to a savage yet uproarious essay on backroom corruption, "Little Tin Box." In a piece of inspired casting, a relative unknown, Tom Bosley, was catapulted to stardom in the title role. The Pulitzer Prize judges felt more than the casting was inspired. For only the third time in its history the award went to a musical. The first musical to win it had been another political satire, *Of Thee I Sing*, the second winner was *South Pacific*. Both were arguably operettas, while *Fiorello!*, despite some operetta-like overtones, was at heart solid musical comedy.

So was *Gypsy*, which won no Pulitzer, though it was not undeserving. Nor could *Gypsy* be said to have taken a roseate view of the worlds of fading vaudeville and sleazy burlesque which were its setting. Nevertheless, as one of its great songs proclaimed, everything came up roses. Gypsy was Gypsy Rose Lee, probably the most celebrated stripper of Minsky's palmiest days, but while the musical was drawn from her autobiography, she commanded center stage only intermittently and fitfully. Looming larger in every way was her pushy, self-serving but self-deceiving stage mother, Rose. The part gave Ethel Merman the juiciest role of her career, a rare opportunity to develop a wide but dramatically legitimate range of emotions and to probe many of them with a depth not typical of the genre. She handled the chore triumphantly. Yet *Gypsy*, a tour de force for Miss Merman, was not a star vehicle. It told a tragicomic story of misguided mother love with honesty and force, giving even its minor figures a flesh-and-blood reality often denied them on musical-comedy stages. At the same time, it never lost sight of the fact that it was a theatre piece, and a musical comedy at that. Sometimes the comedy was implicit, in the charac-

ters' unstated attitudes toward their lives and in the essential absurdities of show business itself. Sometimes it was explicit. But underplayed
or underlined, it was always on the mark.

Jule Styne's melodies and Stephen Sondheim's lyrics took into account the seedy home life, the even seedier back stages, the tinselly,
phony stages themselves, and the ever-present escapist world of wild
daydreams in which the characters moved. As a result the songs moved
tellingly and affectingly from the pensive "Small World" to the tender
"Little Lamb" to the pulsating, optimistic "Together" and "Everything's Coming Up Roses" to the defiant, "You'll Never Get Away
From Me." In between, the show's theatrical settings permitted any
number of tinkly show tunes. For many, however, the most remarkable song in the show came at the end, a half-frightened, half-bellig-
erent soliloquy called "Rose's Turn," in which Rose faces up to the
reasons for and consequences of her behavior. Sondheim, fresh from
the acclaim of *West Side Story*, proved conclusively he had no superior as a lyricist. Not only could he juggle exquisitely the most unexpected tag and inner rhymes, but his ideas and words flowed with a
remarkable colloquial naturalness that often masked his brilliance.
Styne was at his melodic peak. Five years later, he wrote the score for
another musical biography. *Funny Girl* was also a memorable hit,
thanks largely to Barbra Streisand's often hilarious, often touching delineation of Fanny Brice. However, Styne was not then at his peak,
nor did he have Sondheim for a lyricist.

The few sentimental moments that Loesser allowed himself in *Guys
and Dolls* were thrown to the winds eleven years later when he wrote
the songs for *How To Succeed in Business Without Really Trying*. The
closest thing to a love song, "I Believe In You," was sung by the
show's calculating little hero as he admired himself in the mirror.
"Happy To Keep His Dinner Warm," a paean to domesticity, was
quickly reprised in counterpoint against a more cynical second song.
But then a certain sunny cynicism gamboled throughout this two and
a half hour history of a young conniver's carefully chartered rise to the
top. That claw-marked climb had first been recounted in Shepherd
Mead's bestseller. Willie Gilbert and Jack Weinstock had then turned
it into a straight comedy, before collaborating with Abe Burrows in

their efforts to readapt it for the musical stage. Like Loesser they hewed to a totally unromantic line, right down to the very end when the hero is more interested in his sweetheart's suggestion that he might be president of the United States than in learning how much she loves him. A cast of superb farceurs, headed by Rudy Vallee and Robert Morse, gave the merriment a special zest. Like *Fiorello!* before it, *How To Succeed in Business Without Really Trying* garnered a Pulitzer Prize. *How To Succeed's* music was not as endearingly melodic as the music from the rest of these great musical comedies, but the show has been revived with regularity and has not suffered neglect like *One Touch of Venus* and *Lady in the Dark*. Only time will tell whether the show shoots full of holes the proposition that "standards" promote revivals.

The great performers of the era were mixed in their styles and in their loyalty to the theatre. Many of the oldest, most loyal stars said farewell to the footlights during this period from the early 1940s to the early 1960s. Top names such as Ethel Merman, Beatrice Lillie, Bobby Clark, Bert Lahr come to mind at once. Early in this era one superb comedienne, tall, big-eyed Carol Channing, with her "squeak to growl" voice, won over Broadway and retained its affection decade after decade. *Gentlemen Prefer Blondes* and *Hello, Dolly!* were her most popular vehicles. A great young clown in the old tradition, Nancy Walker, left audiences in stitches. *On the Town, Look Ma, I'm Dancin'*, a revue called *Phoenix '55*, and *Do Re Mi* gave her her best opportunities. But she had hard luck finding suitable vehicles and ultimately went west. Sweet, doll-faced Barbara Cook, a fine singer, had better luck. Most of her successes were operettas, but regardless of genre she made *Plain and Fancy, Candide, The Music Man*, and *She Loves Me* all the more enjoyable. Significantly, in a musical theatre increasingly dominated by fine choreographers, two of the most exciting new stars were brilliant dancers—Gwen Verdon and Chita Rivera. Miss Verdon injected her special sparkle into *Can-Can, New Girl in Town, Redhead, Damn Yankees* and *Sweet Charity;* Miss Rivera added hers to *West Side Story,* and *Bye Bye Birdie*. Together they brought a sizzling theatricality to *Chicago*.

Among the men, Alfred Drake was assuredly the best and most glamorous of the newcomers, although he was given a run for his money by Richard Kiley and, briefly, John Raitt. For the most part these men shone best in operetta, although Drake was the original star of *Kiss Me, Kate,* Kiley was the leading man in *No Strings,* and Raitt in *The Pajama Game.* Zero Mostel and the perennially boyish Robert Morse enlivened many a show. Mostel triumphed in *A Funny Thing Happened on the Way to the Forum* and an operetta, *Fiddler on the Roof,* while Morse sent audiences home laughing from *Say Darling, Take Me Along, How To Succeed in Business Without Really Trying, Sugar,* and *So Long, 174th Street.*

Broadway needed its stars, for problems were developing. The failure of Loesser's score for *How to Succeed* to offer one outstanding tune was indicative. At the end of the period, a new style of music swept the Western world. Thereafter, even when Broadway occasionally did produce a masterful score or a capital tune, it rarely received a proper hearing.

13

Rock and Roll;
Pitch and Toss

The 1948–49 theatrical season was a banner year. Although fewer than eighty shows opened on Broadway, the number and quality of superior entertainments and the excitement they generated had probably not been equaled since the glory days of the 1920s. It was a season in which both tradition and experiment flourished. Arthur Miller found pathos and tragedy in the commonplaces of a failed salesman; Maxwell Anderson found poetry and tragedy in the life of Anne Boleyn. Jean Giraudoux turned the lunacies of a Parisian madwoman into a comic rhapsody, while a minor playwright, Robert McEnroe, and a major actor, José Ferrer, combined to bring to life a picaresque vagrant. Fay Kanin set a marvelous drawing-room comedy on a college campus; Moss Hart set one in the suite of a hotel during a play's tryout. And these were merely a few of the best straight plays. Nor did the musical theatre take a back seat. Hits included *Where's Charley:*; *Lend an Ear; Kiss Me, Kate,* and *South Pacific.* Even such failures as *Small Wonder, Magdalena,* and *Love Life* were innovative and filled with memorable goodies.

As *the Girls Go,* a lively, unabashedly old-fashioned song-and-dance show, was brought into the Winter Garden on November 13 by Broadway's most flamboyant producer, Mike Todd. For critics and playgoers its stellar attraction was unquestionably Bobby Clark, up to his roustabout antics in his painted-on glasses. Clark was not alone on stage. Every Todd musical featured a long line of beautiful girls in splashy, often revealing, costumes. They kicked and smirked and posed against a background of equally splashy sets. Since 1948 had been an election year, William Roos's libretto was somewhat topical. It was set four years in the future and recounted the events following the election of America's first woman president. But neither the book nor the lyrics paid serious attention to politics, and what little satire the show offered was almost incidental. Instead, Clark, as "The First Gentleman," cavorted with chorus girls and played a small part in his son's romance with a young lady. Harold Adamson and Jimmy McHugh's songs for the show were delightful, however unadventuresome most of them were. The failure of any to win widespread attention, however, may have contributed to the show's commercial failure. *As the Girls Go* cost a then staggering $340,000 to bring in, and, though it ran a more than creditable 420 performances, it never recouped its initial expenses. It was an ominous portent of things to come on Broadway.

The fate of one Adamson-McHugh song in particular also foreshadowed the future. That song was "Rock, Rock, Rock!," whose sheet music called for a "Groovy" tempo. A pronouncedly rhythmic number, it also coupled a catchy melody and an intelligently developed melodic line with a reasonably bright and literate lyric. Audiences applauded it and just as quickly forgot it. Although a recording strike in 1948 may have deprived it of broader acclaim, as it may have prevented all the show's songs from becoming popular, Broadway was also still ignorant of what later became rock music. As a fresh musical idiom, rock was in its infancy. Not until the mid-fifties did it win national popularity; not until the arrival of the Beatles in the early sixties did it become a national passion.

Until then, Broadway largely ignored rock music, much as it had swing in the late thirties. Broadway audiences were aging. The plumber and his girlfriend for whom George M. Cohan had cavorted had long since deserted to films, so had the secretaries of the 1920s who had scurried from work to Saturday matinees. The youngsters of the thirties were beginning to have children of their own, children for whom live theatre was more often than not a special treat instead of a regular, natural place of entertainment. Broadway was increasingly dependent on middle-aged, expense-account pleasure-seekers. And rock music, certainly more than early ragtime or jazz and possibly even more than swing, was a totally youthful expression.

Rock music's pedigree was complicated, a mesh of the black influences of ragtime, jazz, and swing with white ones, notably western and country motifs. For the most part, the youngsters who developed the new idiom were not, like some earlier musicians, theatre-oriented. Most had been weaned on records, radio, films, and, later, television. Many had never set foot in a legitimate theatre. A second hindrance to rock's quick acceptance by live theatre was that appropriate entrees were no longer available. Revues and black musicals, which had once helped spread the gospel of ragtime and jazz, were both out of vogue in this period. Western and country settings were employed infrequently and were more likely to deal with stories unfolding in times past. Not until 1960 did Broadway face up to the emerging vogue.

At the end of the 1959-60 season, rock and roll's growing attraction to teenagers was the subject of *Bye Bye Birdie*, a musical comedy that was captivatingly musical and delightfully comic but which, despite its story, kept real rock and roll at arm's length. That story (suggested by rock star Elvis Presley's going into the army) told how a rock star's agent and the agent's fiancée conspire to keep the star's popularity alive while he serves his army stint. Their plan is to have their star sing his latest song to a carefully selected, typical American teenager, to surround the event with widespread publicity, and to survive on the song's publicity and sales of the music until he is discharged. The confrontation of show business types with middle Americans provided much of the fun. A superb book by Michael Stewart and score by Charles Strouse and excellent performances were tied together by Gower Champion's humorous, inventive staging. *Bye Bye Birdie* was a smash hit, even if its few, brief, satiric looks at rock music were among the show's minor moments.

Rock really did not take hold on Broadway until the 1967–68 season, by which time it dominated popular American music every-where else. On radio, the old "Hit Parade" was long gone, but if it had remained it would long since have forgotten the theatre. Theatre music, still largely traditional, had been shoved aside rudely by rock and its offspring.

Broadway did not embrace rock brazenly. Two musicals that did win true acceptance—*Hair* and *Your Own Thing*—both started in tiny playhouses far from Times Square. In fact, *Your Own Thing* never left its off-Broadway home, although road companies played long runs at traditional legitimate houses across the country. The two musicals took sharply different tacks, suggesting early on how widely rock music was to be adapted to mainstream requirements in the theatre. *Your Own Thing*, based on Shakespeare's *Twelfth Night*, was reasonably well plotted, witty, and irreverent. *Hair*, on the other hand, was little more than a loose-jointed, uncertain chronicle, preachy and rebellious. But whereas *Your Own Thing*'s music was lackluster, *Hair*'s was gor-geously and memorably melodic. *Your Own Thing* suggested that rock could be adapted readily to much traditional libretto fare; *Hair* dem-onstrated that rock could bring people to sing Broadway songs again. Unfortunately, time was to demonstrate that librettists for rock shows had little conception of theatrical requirements. Nor did lyricists or composers. They hardly ever provided a simple, immediately singable love song and simple yet sophisticated melodies to other songs that could be applied to superior lyrics. Broadway's rock composers were to turn out largely forgettable scores.

Despite the success of *Hair* and *Your Own Thing*, few rock musi-cals were presented and most failed. Although they displayed a sur-prising range of subject, they quickly developed an identifiable set of characteristics. Critics had long since abandoned the cry that musical comedies had no plot. Yet, excepting those shows based on older plays or well-plotted stories, rock musicals sometimes went a step farther and had no real libretto, at least not in any accepted guise—that is, the major developments in their stories were not revealed by stretches of dialogue. These musicals were entirely sung. For that reason, no doubt, *Jesus Christ Superstar* and a few other rock musicals were ad-vertised as rock operas. A fair description, although some classical opera aficionados might shudder. Even the seemingly more traditional librettos often appeared to be merely the flimsiest frames for musical numbers. As often as not, stories were little more than outlines and characters were cardboard personifications. *Grease*, for example, was set in and around Rydall High, where boys meet girls, lose them, and win them back. But the story was forever taking a back seat to the show's affectionate look at the foibles and music of the 1950s. *Hair's* ostensible hero is a young man named Claude who debates whether or not to burn his draft card in protest of the Viet Nam War. In the end he joins the army and is killed. But for most of the show Claude is lost in a ragtag, eddying montage of life and protest in the 1960s.

The dialogue of these shows was determinedly slangy, frequently obscene. It was a far cry from the careful dialogue that even the toughest earlier shows had offered. And the language spilled into the lyrics, which in many ways were even less traditional than the dialogue. Most of the time they wallowed in banality. Gone was most of the clever wordplay, the imaginative rhyming, the witty tinkering with ideas. Deadening commonplaces, irritating repetitions, and blatant preaching became the rule, with few exceptions.

The music these words were set to was not the purest rock, no more than theatrical ragtime or theatrical jazz represented unadulterated expressions of those idioms. But whereas Broadway ragtime and Broadway jazz had quickly become syntheses that were effective on stage and off and that left behind a catalogue of enduring songs, theatrical rock has remained a pallid stepchild of the best commercial rock. True, *Hair* offered "Aquarius" and "Good Morning, Starshine," and *Jesus Christ Superstar* gave us its title song and "I Don't Know How To Love Him." For the most part, however, rock musicals have produced few songs that promise to survive. At best many of them have presented a single song that has become identified with the show, much as a theme song once identified a big band. But these songs probably have little more durability than did most big band themes after the band had been dispersed.

Rock musicals certainly abetted the spread of one horrendous Broadway practice, amplification. Much modern rock seems to have been created for electronic instruments and sounds best when played on them. But live theatre retreats a giant step from its audiences once it is amplified.

Let's look at one of the most successful American rock musicals, *The Wiz*. The show differed from other rock musicals in its use of a well-plotted story, in its avoidance of propagandizing, in its wholesomeness and therefore its appeal to children, and in the lavishness of its physical production. These elements may have accounted for the show's success.

The Wiz was a modern-day retelling of L. Frank Baum's *The Wizard of Oz*, whose story first had been adapted for Broadway in 1903 (starring Fred Stone and Dave Montgomery) and had later inspired the beloved MGM film. The 1975 version had an entirely black cast, and unlike many earlier black shows that had been written by whites, it was also created by blacks. The songs were based on rhythm and blues and featured the insistent, quite heavy beat of soul music, a black derivative of rock and roll. Almost as soon as the curtain was up and a brief glimpse of a Kansas farmhouse had set the scene, the tornado swept little Dorothy into her neverland. In the original 1903 stage version, curtain after curtain had opened revealing "twenty-seven heavens and nine hells of scenery." Advertisements had warned 1903 ticket-holders not to miss this five minutes of theatrical trickery. In the 1975 production, a ballet (a rarity in rock musicals) transports Dorothy and her audiences, the tornado personified by a dancer in clouds of dark drapery. Once on the yellow brick road, Dorothy meets up with a scarecrow, a tinman, and a lion, as she did in the book and film.

The meeting of Dorothy and the scarecrow led into the singing of *The Wiz*'s most popular song, "Ease On Down The Road." The song was catchier than most Broadway rock numbers but was still typical in that its basically repetitive melody was played out to a please-pick-me-up beat and sung to simple, repetitive lyrics. The phrase "ease on down" was repeated five times in a row at one point. "Who Do You Think You Are?" repeated its title six times running. In another song, "Home," a whole stanza went unrhymed. Yet the lyrics did have a naturalness and were free of involutions and artificial expressions that were a not uncommon drawback of older lyrics. Still, they lacked the elegance and invention of the best of the old school, as did the music that accompanied them.

In the early 1970s, two other categories of musicals appeared on Broadway along with rock musicals. One was the director's musical, the other the "conceptualized" musical. They were close kin. More often than not the director of the director's musical was a trained choreographer. One can see the origins of this school in the early works of Jerome Robbins—*On the Town* and *West Side Story*. The choreography was an integral part of both shows. Sadly, by the late 1960s Robbins had tired of and left Broadway. His mantle fell to a number of inspired successors including Bob Fosse, Gower Champion, and Michael Bennett.

Choreographers had directed shows in the past. Ned Wayburn, who is reputed to have invented tap dancing, did double duty for many musicals in the first decade of the century. But the shape and spirit of those musicals had been predetermined. Wayburn and his ilk undoubtedly asked for and got many changes, but essentially they were simply the original pacemakers. Of course, at the time, that was all even the best director-choreographers could do with many of the shows. They might invent imaginative moments of staging and impart a distinctive style, but they could do little else.

Increasingly, however, musicals came to originate in choreographer's fertile visions. Bob Fosse's stamp was evident in his earliest efforts such as *The Pajama Game* in 1954 and *Damn Yankees* in 1955, shows for which he did only the choreography. His talent began to reach full bloom in vehicles he helped to create for his wife at the time, Gwen Verdon. *New Girl in Town* (in which Fosse again handled only the choreography) and *Redhead*, in fact, had little of superior merit except his choreography and Miss Verdon's dancing. With *Sweet Charity*, *Pippin*, and *Chicago* Fosse hit full stride. Although *Sweet Charity* had a hit-laden score by Cy Coleman, Dorothy Fields's wonderful lyrics, and, of course, Miss Verdon, the show was really Fosse's. Working with lesser material in *Pippin* and *Chicago*, only his own special razzle-dazzle turned those shows into unforgettable theatre. His eccentric, angular movements, his tight-knit ensembles that sometimes exploded dramatically and unexpectedly, all seemingly derived from street-dancing, became Fosse trademarks. His peak may well have been reached in his 1978 choreographic revue, *Dancin'*.

Gower Champion was never offered or perhaps never wanted the free hand Fosse was given. His later work was with sturdily constructed book shows set in past times. He imbued each with a lighthearted sense of period elegance and apparently suggested many of the clever theatrical gimmicks that punctuated his staging. *Hello, Dolly!*, *I Do! I Do!*, and *Sugar* probably would have enjoyed long runs even without his guidance, but his imprint made them all the more graceful and memorable. Champion's career came to a theatrical end when he died on the eve of the 1980 premiere of his last show, *42nd Street*, a stage version of the old film.

Michael Bennett's style is less readily categorized. He seems the most eclectic and freewheeling of the contemporary masters. Bennett directed *Seesaw* and co-directed *Follies* but came fully into his own with *A Chorus Line* and the failed *Ballroom*. Both, as their titles hint, were preeminently dance shows.

A Chorus Line began life in 1975 as an experimental offering on one of Joseph Papp's innovative stages, as had *Hair* and a rock version of *Two Gentlemen of Verona*. Bennett, working with librettists James Kirkwood and Nicholas Dante, composer Marvin Hamlisch, and lyricist Edward Kleban, developed the show from his own and other dancers' personal experiences in auditioning. Performed without scenery except a mirrored rear wall, and, but for one scene, with the entire cast in work clothes, the show followed a group of aspirants from auditions to selection. There was hardly any plot as such. Instead, the show consisted largely of a series of vignettes that allowed each applicant to give some of his or her history and dreams. Apart from "One," a song sung during the show and reprised for an effective finale, and possibly "What I Did For Love," the songs had no life outside the show. Even "One" was really little more than the musical's recognizable theme. Hamlisch's music was thoroughly contemporary, filled with modern coloring, and often throbbing with rhythm but was not readily embraceable. The lyrics were hardly June-moon, touching on homosexuality, suicide, and similar delicate subjects, and occasionally resorting to offensive expressions. Nonetheless, as a piece of theatre, as a metaphor in stage terms for the frightening battle for acceptance, A *Chorus Line* was spellbinding.

In a sense, these director's shows were a throwback to the sort of musicals written before the advent of the great composers, lyricists, and librettists. As such, they demonstrated that something was amiss or lacking in our modern lyric theatre. They were not, with very few exceptions, vehicles for star performers—vehicles with indifferent librettos and songs that stars held together, dominated, and made palatable. But they were, nonetheless, star vehicles: the star was the director-choreographer. Texture often seemed more important than text. Luckily, men such as Fosse, Champion, and Bennett were theatrical artists of the highest caliber. Their shows always had a distinctive sense of style and tone, an artistry absent in most of the older vehicles.

Text, however, was paramount in the conceptualized musical. Although the best director's musicals evolved out of thoughtful concepts, their theatrical tension and electricity came from the director's rather than the writer's genius. Conceptualized musical comedy is best exemplified by the works of Stephen Sondheim. Indeed, it can be argued that Sondheim developed the very idea and remains alone in working it out with flair. Sondheim had delighted Broadway with both operettas (A *Little Night Music* and *Sweeney Todd*) and musical comedies, although, curiously, he seems to have taken opposite tacks in approaching the two genres. His operettas were almost a throwback to early comic operas, despite striking superficial departures. They were more operatic, more pervasively lyrical than many modern operettas, and their underlying misanthropy gave them a comic perspective not unlike Gilbert's, if more publicly jaundiced. Sondheim's musical comedies—*Company* in 1970 and *Follies* in 1971—were equally venturesome, but they were also forward-looking. Officially, he created only the songs, but his stamp, along with Hal Prince's sympathetic, dynamic direction, imbued each show with his imagination, style, and craftsmanship.

Of course, Sondheim had served an apprenticeship, albeit a singularly amazing and distinguished one, before he created these fully realized pieces. In the late 1950s his lyrics for *West Side Story* and

Gypsy instantly had revealed his word witchery. In 1962, *A Funny Thing Happened on the Way to the Forum* suggested his gift for melody might match his ability at rhyming. Two years later, an imaginative but failed effort, *Anyone Can Whistle*, pointed to the venturesome road Sondheim was to take. After writing lyrics for Richard Rodgers's *Do I Hear a Waltz?*, a failure, Sondheim turned his attention to *Company*.

Stanley Green has noted in *The World of Musical Comedy*, "Though the story was originally written as a play without songs, *Company* developed into a tightly coordinated 'conceptual' musical, utilizing music and dance not only as part of the action but also to comment on the actions and characters. It was particularly applauded for the way director Prince and composer-lyricist Sondheim fused their talents to achieve a seemingly seamless production that was almost as much a revue with a theme as it was a book musical." And while the theme was an old musical standby—marriage—*Company* turned its back on the standard, simplistic, and saccharine boy-meets-girl approach.

Sondheim's hero was Bobby, a thirty-five-year-old bachelor. His married friends are curiously eager to marry him off—even though their own marriages are far from happy and some are even on the rocks. When the men are candid with Bobby, they admit to mixed emotions about their marriages, confessing they are "Sorry-Grateful." Bobby's freedom and his dreams about his girl friends lead the men to respond "You Could Drive A Person Crazy," and they insist "Have I Got A Girl For You." While not yet prepared to take the plunge, Bobby is certain "Someone Is Waiting." His evenings with a wacky intellectual and with an airline stewardess, however, are unrewarding. For all their disagreements Bobby and his friends enjoy being "Side By Side By Side" and ask each other "What Would We Do Without You?" Time passes, but neither Bobby's life nor those of his friends change all that much. One of the married women sums up her comfortable if boring existence in "The Ladies Who Lunch" (admirably performed in the original production by Elaine Stritch). In the end, Bobby does seem ready to find a wife, but Sondheim's closing lyric leaves a modicum of doubt about that readiness. George Furth's dialogue was superb—observant, comic, and literate. Sondheim's lyrics were even better, a stunning *tour de force*.

The even freer form of *Follies* centered on a reunion of old performers who had once played in the elaborate revues of a Ziegfeld-like producer. It took a tough view of a bygone Broadway, but essentially, like *Company*, it was an examination of the city-living attitudes of the day. The theatre, after all, is essentially a city phenomenon.

Like *Lady in the Dark* and *One Touch of Venus*, two equally brilliant and innovative musicals, *Company* and *Follies* are not given major revivals as so many other musicals of their period. Sondheim's music, while thoughtful and appropriate, has won little popularity away from the theatre. The failure to provide blockbuster songs seems to continue to keep more deserving musicals from becoming a part of the standard revival repertory.

Not even several superior songs Sondheim wrote for *Merrily We Roll Along* allowed the musical to survive. Once more the songs suggest that Sondheim's talents, however prodigious, are not the sort that can readily be called lovable. In reviewing *Merrily We Roll Along*, Frank Rich of the *New York Times* suggested Sondheim's songs "can tear through us with an emotional force" and can "hurt." Rich compared Sondheim's ability to Gershwin's, ignoring the fact that Gershwin's great songs, although at times pugnaciously assertive, were on a more colloquial level and were demandingly memorable. Moreover, the musicals for which Gershwin created his songs were usually far more formal and amiable. Even biting satires such as *Strike Up the Band* and *Of Thee I Sing* sunnily told full-fledged, fundamentally traditional musical-comedy stories. (When bitterness got the better of Gershwin's librettists, as in *Let 'Em Eat Cake*, the result was a short-lived failure.)

Sondheim's artful writing and Hal Prince's equally artful staging cannot override the loosely plotted books they work with. Their librettos do not offer playgoers neat, attractively wrapped packages, stories with beginnings, middles, and ends. Nor are the contents of those packages a pleasant surprise. There is an underlying misanthropy and disillusionment permeating virtually everything Sondheim and his associates have put their hand to. A Sondheim show is rarely comfortable or comforting.

Michael Bennett's shows do have a comfortable feeling, despite the unpleasantness they contain, perhaps simply because their stories have a more obvious formality—or, at least, form. In the 1981 hit *Dreamgirls*, Tom Eyen's tale of the hard road from rags to riches in show business is somewhat clichéd, but the clichés are relaxing and soothing, a far cry from the sharply observant books and lyrics that Sondheim and his associates sometimes have made seem like disturbing sociological studies. Similarly, Henry Krieger's music, while it might lack the intellectual appeal of Sondheim's best, is, as one critic noted, instantly "hummable." For all the sordidness Bennett throws at his audience, he clearly understands its need for some pleasant certainties. His direction and choreography for *Dreamgirls* also played a major part in the show's success.

Where is our musical comedy, then? Artistically, we have come a long way from the crudities of *The Brook, The Umpire, The Hen Pecks*, and even *Manhattan Mary*. For all their obnoxious obscenities, our best librettos are now literate and adult, as are our best lyrics. Musically, the answer may not be as satisfying. Take any season twenty-five or fifty years ago—indeed, just about any season from World War I to the onset of the Vietnam War—and think of all the wonderful songs that came from the musical comedies of that season. How many of them are still sung and loved! What song from any musical in the past ten years has won widespread popularity and gives any promise of enduring (aside from "Tomorrow")? The question is a bit unfair. There have been a handful of excellent songs from musical comedies, especially those by Cy Coleman, that have been denied the fame they deserve because of the rage for rock and roll and the unwillingness of radio and television to promote theatre music the way they once did. But Broadway has also been somehow unable to respond to rock and roll the way it responded earlier to ragtime and to jazz. It has held to largely traditional musical styles—idioms that may be exhausted, though it is doubtful. The emphasis on composers in earlier shows has now passed to librettists, directors, and choreographers. Sondheim excepted (and even his dominating brilliance lies in his lyrics) we no longer think of a Gershwin show, a Porter show, or a Styne show.

We are instead implicitly asked to consider a show as a totality in which the perceptive director has achieved the careful integration of song and story that the authors have aimed for. But the integration of song and story goes much farther back than ardent modernists will admit. *Oh, Boy!*, for example, was a largely integrated musical comedy, whose humor attempted to derive from the characters and situations and whose songs helped to forward the action. The integration was not perfect, but neither is it in, say, *A Chorus Line*. Just as almost any song in *Oh, Boy!* might have been replaced by another song expressing different sentiments at a different point in the story, so the characters in *A Chorus Line* could have been given different songs telling different histories. If integration suggests a certain inevitability, then it will probably always remain a persistent goal at its most ideal. Each era will see its best musical comedies reaching for it with hope.

No doubt many playgoers and critics would say "good riddance" to what they perceive as older musical-comedy mannerisms. They would argue that the long runs of *Annie*, *A Chorus Line*, and other recent musical comedies prove the public is getting what it wants. But is that theatre-hungry public merely taking the best of what it gets, as the counter-argument runs? Moreover, the longest runs of a bygone era were regularly the inane Hippodrome spectacles, and the brilliant early Gershwin and early Rodgers and Hart musical comedies did not run as long as some of the less artful competition. Thus, while long runs do indicate public acceptance, they do not really establish the value of a work, which becomes apparent with the passing of time. How would the public respond to a topflight musical comedy with an easy-going, happy story and a batch of instantly singable songs?

In short, though our best directors, choreographers, librettists, and lyricists are creating noteworthy works, one can only wonder if something is not lacking in them. Mature, probing musicals are fine; our theatre would be poorer without them. But musical comedy simply is no longer as insinuatingly musical or as ingratiatingly comic as it used to be. Surely popular melody and lighthearted humor need not be artless. Just as surely, Broadway is big and booming enough to have room for musical comedies that are musical and comical.

Appendix 1

Principals and Credits
for Important Musical Comedies

Evangeline. Music by E. E. Rice; libretto by J. Cheever Goodwin. Niblo's Garden. July 27, 1874. Ione Burke (Evangeline), Connie Thompson (Gabriel), J. W. Thompson (Lone Fisherman), Louis Mestayer (Catherine). Principal songs: *"Sweet Evangeline," "Golden Chains," "Heifer Dance," "My Best Beloved."* 16 *performances.*

The Mulligan Guards' Ball. Music by David Braham; libretto by Edward Harrigan. Theatre Comique. January 13, 1879. Edward Harrigan (Dan Mulligan), Tony Hart (Tommy Mulligan), Annie Yeamans (Cordelia Mulligan). Principal songs: "The Mulligan Guard," "The Babies On Our Block." Approximately 120 performances.

Cordelia's Aspirations. Music by David Braham; libretto by Edward Harrigan. Theatre Comique. November 5, 1883. Edward Harrigan (Dan Mulligan), Tony Hart (Rebecca Allup), Annie Yeamans (Cordelia Mulligan). Principal songs: "My Dad's Dinner Pail," "Just Across From Jersey," "Sam Johnson's Cake Walk." Approximately 200 performances.

Adonis. Music borrowed from various composers; libretto by William F. Gill. Bijou Theatre. September 4, 1884. Henry E. Dixey (Adonis), Amelia Summerville (Rosetta), Jennie Reiffarth (Duchess), Louise V. Essing (Artea), Herbert Gresham (Marquis), George Howard (Bunion). Principal songs: "I'm English You Know," "A Very Susceptible Statue." 603 performances.

A Trip to Chinatown. Music by Percy Gaunt; book and lyrics by Charles H. Hoyt. Madison Square Theatre. November 9, 1891. Harry Conor (Welland Strong), George A. Beane, Jr. (Ben Gay), Anna Boyd (Mrs. Guyer). Principal songs: "The Bowery," "Reuben And Cynthia," "After The Ball." 657 performances.

A Gaiety Girl. Music by Sidney Jones; book by Owen Hall; lyrics by Harry Greenbank. Daly's Theatre. September 18, 1894. Charles Ryley (Capt. Goldfield), Maud Hobson (Virginia Forrest), Blanche Massey (Alma Somerset). Principal songs: "Private Tommy Jones," "Sunshine Above." 79 performances.

The Belle of New York. Music by Gustave Kerker; book and lyrics by Hugh Morton. Casino Theatre. September 28, 1897. Edna May (Violet Gray), Harry Davenport (Harry Brown), Dan Daly (Ichabod Brown). Principal songs: "They All Follow Me," "The Purity Brigade." 56 performances.

The Sultan of Sulu. Music by Alfred G. Wathall; book and lyrics by George Ade. Wallack's Theatre. December 29, 1902. Frank Moulan (Ki-Ram), Blanche Chapman (Pamela Jackson). Principal songs: "Since First I Met You," "The Smiling Isle." 192 performances.

Little Johnny Jones. Music, book, and lyrics by George M. Cohan. Liberty Theatre. November 7, 1904. George M. Cohan (Johnny), Ethel Levey (Goldie Gates), Jerry Cohan (Anthony Anstey). Principal songs: "The Yankee Doodle Boy," "Give My Regards To Broadway," "Life's A Funny Proposition After All." 52 performances.

Forty-Five Minutes from Broadway. Music, book, and lyrics by George M. Cohan. New Amsterdam Theatre. January 1, 1906. Victor Moore (Kid Burns), Fay Templeton (Mary), Donald Brian (Tom). Principal songs: "Forty-five Minutes From Broadway," "Mary's A Grand Old Name," "So Long, Mary." 90 performances.

George Washington, Jr. Music, book, and lyrics by George M. Cohan. Herald Square Theatre. February 12, 1906. George M. Cohan (George Belgrave), Jerry Cohan (Senator Belgrave), Ethel Levey (Dolly Johnson). Principal songs: "You're A Grand Old Flag," "I Was Born In Virginia." 81 performances.

The Red Mill. Music by Victor Herbert; book and lyrics by Henry Blossom. Knickerbocker Theatre. September 24, 1906. Fred Stone (Con Kidder), Dave Montgomery (Kid Conner), Augusta Greenleaf (Gretchen), Joseph M. Ratliff (Doris Van Damm). Principal songs: "The Streets Of New York," "The Isle Of Our Dreams," "Moonbeams," "Every Day Is Ladies' Day With Me," "Because You're You." 274 performances.

Madame Sherry. Music by Karl Hoschna; book and lyrics by Otto Harbach. New Amsterdam Theatre. August 30, 1910. Lina Abarbanell (Yvonne), Ralph Herz (Uncle Theophilus), Jack Gardner (Edward Sherry), Elizabeth Murray (Catharine). Principal songs: "Every Little Movement," "The Birth Of Passion," "The Smile She Means For You," "Put Your Arms Around Me, Honey." 231 performances.

Very Good Eddie. Music by Jerome Kern; book by Guy Bolton and Philip Bartholomae; lyrics by various writers. Princess Theatre. December 23, 1915. Ernest Truex (Eddie), Alice Dovey (Elsie Darling), Oscar Shaw (Dick Rivers). Principal songs: "Babes In The Wood," "Some Sort Of Somebody," "Size Thirteen Collar." 341 performances.

Oh, Boy! Music by Jerome Kern; book by Guy Bolton and P. G. Wodehouse; lyrics by P. G. Wodehouse. Princess Theatre. February 20, 1917. Tom Powers (George Budd), Marie Carroll (Lou Ellen), Anna Wheaton (Jackie), Hal Forde (Jim). Principal songs: "Till The Clouds Roll By," "Nesting Time," "A Pal Like You." 463 performances.

Irene. Music by Harry Tierney; book by James Montgomery; lyrics by Joe McCarthy. Vanderbilt Theatre. November 18, 1919. Edith Day (Irene), Walter Regan (Donald Marshall). Principal songs: "Alice Blue Gown," "Irene." 670 performances.

Mary. Music by Louis Hirsch; book and lyrics by Otto Harbach and Frank Mandel. Knickerbocker Theatre. October 18, 1920. Janet Velie (Mary), Jack McGowan (Jack). Principal songs: "The Love Nest," "Mary," "We'll Have A Wonderful Party." 219 performances.

Sally. Music by Jerome Kern; book by Guy Bolton; lyrics mostly by Clifford Grey. New Amsterdam Theatre. December 21, 1920. Marilyn Miller (Sally), Leon Errol (Duke Constantine), Walter Catlett (Otis Hopper), Irving Fisher (Blair Farquar). Principal songs: "Look For The Silver Lining," "Wild Rose," "Whip-Poor-Will." 570 performances.

Lady, Be Good! Music by George Gershwin; book by Guy Bolton and Fred Thompson; lyrics by Ira Gershwin. Liberty Theatre. December 1, 1924. Fred Astaire (Dick Trevor), Adele Astaire (Susie Trevor), Walter Catlett ("Watty" Watkins), Alan Edwards (Jack Robinson). Principal songs: "Oh, Lady Be Good!," "So Am I," "Fascinating Rhythm." 330 performances.

No, No, Nanette. Music by Vincent Youmans; book by Otto Harbach and Frank Mandel; lyrics mostly by Irving Caesar. Globe Theatre. September 16, 1925. Louise Groody (Nanette), Charles Winninger (Jimmy Smith), Jack Barker (Tom). Principal songs: "Tea For Two," "I Want To Be Happy." 321 performances.

Dearest Enemy. Music by Richard Rodgers; book by Herbert Fields; lyrics by Lorenz Hart. Knickerbocker Theatre. September 18, 1925. Helen Ford (Betsy Burke), Charles Purcell (Capt. Sir John Copeland). Principal songs: "Here In My Arms," "Bye And Bye." 286 performances.

Sunny. Music by Jerome Kern; book and lyrics by Otto Harbach and Oscar Hammerstein II. New Amsterdam Theatre. September 22, 1925. Marilyn Miller (Sunny), Paul Frawley (Tom), Jack Donahue (Jim), Clifton Webb (Harold), Mary Hay ("Weeny"), Joseph Cawthorn (Siegfried). Principal songs: "Who?," "Sunny," "D'ya Love Me?" 517 performances.

Oh, Kay! Music by George Gershwin; book by Guy Bolton and P. G. Wodehouse; lyrics by Ira Gershwin. Imperial Theatre. November 8, 1926. Gertrude Lawrence (Kay), Victor Moore ("Shorty" McGee), Oscar Shaw (Jimmy Winter). Principal songs: "Someone To Watch Over Me," "Maybe," "Do, Do, Do," "Clap Yo' Hands." 256 performances.

Hit the Deck! Music by Vincent Youmans; book by Herbert Fields; lyrics by Clifford Grey and Leo Robin. Belasco Theatre. April 25, 1927. Louise Groody (LouLou), Charles King (Bilge), Stella Mayhew (Lavinia). Principal songs: "Hallelujah!," "Sometimes I'm Happy." 352 performances.

Good News. Music by Ray Henderson; book by Laurence Schwab and B. G. De Sylva; lyrics by B. G. De Sylva and Lew Brown. 46th St. Theatre. September 6, 1927. Mary Lawlor (Connie Lane), John Price Jones (Tom Marlowe), Zelma O'Neal (Flo). Principal songs: "The Best Things In Life Are Free," "Lucky In Love," "Just Imagine," "Good News," "Varsity Drag." 557 performances.

A Connecticut Yankee. Music by Richard Rodgers; book by Herbert Fields; lyrics by Lorenz Hart. Vanderbilt Theatre. November 3, 1927. William Gaxton (Martin), Constance Carpenter (Alice Carter), Nana Bryant (Fay Morgan). "My Heart Stood Still," "Thou Swell," "On A Desert Island With Thee." 418 performances.

Funny Face. Music by George Gershwin; book by Fred Thompson and Paul Gerard Smith; lyrics by Ira Gershwin. Alvin Theatre. November 22, 1927. Fred Astaire (Jimmie Reeve), Adele Astaire (Frankie), Victor Moore (Herbert), William Kent (Dugsie), Allen Kearns (Peter Thurston). Principal songs: "'S Wonderful," "Funny Face," "The Babbitt And The Bromide." 250 performances.

Paris. Music and lyrics mostly by Cole Porter; book by Martin Brown. Music Box Theatre. October 8, 1928. Irene Bordoni (Vivienne Roland), Eric Kalkhurst (Andrew Sabot). Principal song: "Let's Do It." 195 performances.

Fifty Million Frenchmen. Music and lyrics by Cole Porter; book by Herbert Fields. Lyric Theatre. November 27, 1929. William Gaxton (Peter Forbes), Genevieve Tobin (Looloo Carroll), Evelyn Hoey (May De Vere). Principal songs: "You Do Something To Me," "You've Got That Thing," "Find Me A Primitive Man." 254 performances.

The New Yorkers. Music and lyrics by Cole Porter; book by Herbert Fields. Broadway Theatre. December 8, 1930. Charles King (Al), Hope Williams (Alice), Jimmy Durante (Jimmie), Ann Pennington (Lola), Richard Carle (Dr. Wentworth), Marie Cahill (Mrs. Wentworth). Principal songs: "Love For Sale," "I Happen To Like New York." 168 performances.

Girl Crazy. Music by George Gershwin; book by Guy Bolton and John McGowan; lyrics by Ira Gershwin. Alvin Theatre. October 14, 1931. Allen Kearns (Danny Churchill), Willie Howard (Gieber Goldfarb), Ginger Rogers (Molly Gray), Ethel Merman (Kate Fothergill). Principal songs: "Bidin' My Time," "But Not For Me," "Could You Use Me," "Embraceable You," "I Got Rhythm." 272 performances.

Face the Music. Music and lyrics by Irving Berlin; book by Moss Hart. New Amsterdam Theatre. February 17, 1932. Mary Boland (Mrs. Meshbesher), Andrew Tombes (Hal Reisman). Principal songs: "Let's Have Another Cup O' Coffee," "Soft Lights And Sweet Music." 165 performances.

Roberta. Music by Jerome Kern; book and lyrics by Otto Harbach. New Amsterdam Theatre. November 18, 1933. Fay Templeton (Aunt Minnie), Tamara (Stephanie), Helen Gray (Sophie), Ray Middleton (John), Bob Hope

(Huckleberry Haines). Principal songs: "Smoke Gets In Your Eyes," "Yesterdays," "The Touch Of Your Hand," "You're Devastating." 295 performances.

Anything Goes. Music and lyrics by Cole Porter; book by Guy Bolton, P. G. Wodehouse, Howard Lindsay, and Russell Crouse. Alvin Theatre. November 21, 1934. Ethel Merman (Reno Sweeney), William Gaxton (Billy Crocker), Victor Moore (the Rev. Dr. Moon), Bettina Hall (Hope Harcourt). Principal songs: "Anything Goes," "All Through The Night," "I Get A Kick Out Of You," "You're The Top," "Blow, Gabriel, Blow." 420 performances.

On Your Toes. Music by Richard Rodgers; book by Richard Rodgers, Lorenz Hart, and George Abbott; lyrics by Lorenz Hart. Imperial Theatre. April 11, 1936. Ray Bolger (Junior), Doris Carson (Frankie), Luella Gear (Peggy), Monty Woolley (Sergei). Principal songs: "There's A Small Hotel," "Slaughter On Tenth Avenue" (ballet). 315 performances.

The Boys from Syracuse. Music by Richard Rodgers; book by George Abbott; lyrics by Lorenz Hart. Alvin Theatre. November 23, 1938. Teddy Hart and Jimmy Savo (Dromio), Eddie Albert and Ronald Graham (Antipholus), Muriel Angeles (Adriana), Marcy Westcott (Luciana), Wynn Murray (Luce). Principal songs: "Falling In Love With Love," "This Can't Be Love." 235 performances.

Cabin in the Sky. Music by Vernon Duke; book by Lynn Root; lyrics by John Latouche. Martin Beck Theatre. October 25, 1940. Ethel Waters (Petunia Jackson), Dooley Wilson (Little Joe), Rex Ingram (Lucifer, Jr.), Todd Duncan (The Lawd's General). Principal songs: "Taking A Chance On Love," "Cabin In The Sky," "Honey In The Honeycomb." 156 performances.

Pal Joey. Music by Richard Rodgers; book by John O'Hara; lyrics by Lorenz Hart. Ethel Barrymore Theatre. December 25, 1940. Vivienne Segal (Vera), Gene Kelly (Joey). Principal songs: "Bewitched," "I Could Write A Book." 374 performances.

Lady in the Dark. Music by Kurt Weill; book by Moss Hart; lyrics by Ira Gershwin. Alvin Theatre. January 23, 1941. Gertrude Lawrence (Liza Elliott), Bert Lytell (Kendall Nesbitt), Macdonald Carey (Charlie Johnson), Victor Mature (Randy Curtis), Danny Kaye (Randall Paxton). Principal songs: "My Ship," "The Saga Of Jenny." 467 performances.

One Touch of Venus. Music by Kurt Weill; book by S. J. Perelman and Ogden Nash; lyrics by Ogden Nash. Imperial Theatre. October 7, 1943. Mary

Martin (Venus), Kenny Baker (Rodney Hatch), John Boles (Whitelaw Savory). Principal songs: "Speak Low," "That's Him," "The Trouble With Women." 567 performances.

Mexican Hayride. Music and lyrics by Cole Porter; book by Herbert and Dorothy Fields. Winter Garden Theatre. January 28, 1944. Bobby Clark (Joe Bascom), June Havoc (Montana), Wilbur Evans (David Winthrop). Principal songs: "I Love You," "Sing To Me, Guitar," "There Must Be Someone For Me." 481 performances.

On the Town. Music by Leonard Bernstein; book and lyrics by Betty Comden and Adolph Green. Adelphi Theatre. December 28, 1944. Betty Comden (Claire de Loon), Nancy Walker (Hildy), Sono Osato ("Miss Turnstiles"), Adolph Green (Ozzie), John Battles (Gaby). Principal songs: "Lucky To Be Me," "New York, New York." 463 performances.

Annie Get Your Gun. Music and lyrics by Irving Berlin; book by Herbert and Dorothy Fields. Imperial Theatre. May 16, 1946. Ethel Merman (Annie Oakley), Ray Middleton (Frank Butler). Principal songs: "Doin' What Comes Natur'lly," "They Say It's Wonderful," "The Girl That I Marry," "There's No Business Like Show Business," "I Got The Sun In The Morning." 1,147 performances.

Finian's Rainbow. Music by Burton Lane; book by E. Y. Harburg and Fred Saidy; lyrics by E. Y. Harburg. 46th St. Theatre. January 10, 1947. Ella Logan (Sharon), Albert Sharpe (Finian), David Wayne (Og). Principal songs: "How Are Things In Glocca Morra?," "Look To The Rainbow," "If This Isn't Love." 725 performances.

Where's Charley? Music and lyrics by Frank Loesser; book by George Abbott. St. James Theatre. October 11, 1948. Ray Bolger (Charley), Byron Palmer (Jack), Doretta Morrow (Kitty), Allyn Ann McLerie (Amy). Principal songs: "My Darling, My Darling," "Once In Love With Amy." 792 performances.

Kiss Me, Kate. Music and lyrics by Cole Porter; book by Samuel and Bella Spewack. New Century Theatre. December 30, 1948. Alfred Drake (Fred), Patricia Morison (Lilli), Lisa Kirk (Lois), Harold Lang (Bill). Principal songs: "So In Love," "Wunderbar," "Always True To You In My Fashion," "Brush Up Your Shakespeare." 1,077 performances.

Guys and Dolls. Music and lyrics by Frank Loesser; book by Abe Burrows and Jo Swerling. 46th St. Theatre. November 24, 1950. Robert Alda (Sky

Masterson), Vivian Blaine (Adelaide), Sam Levene (Nathan Detroit), Isabel
Bigley (Sister Sarah). Principal songs: "A Bushel And A Peck," If I Were A
Bell," "I'll Know," "Adelaide's Lament." 1,200 performances.

The Pajama Game. Music and lyrics by Richard Adler and Jerry Ross; book
by George Abbott and Richard Bissell. St. James Theatre. May 13, 1954.
John Raitt (Sid), Janis Paige (Babe), Eddie Foy, Jr. (Hines), Carol Haney
(Gladys). Principal songs: "Hey, There," "Hernando's Hideaway," "Steam
Heat." 1,063 performances.

Damn Yankees. Music and lyrics by Richard Adler and Jerry Ross; book by
George Abbott and Douglass Wallop. 46th St. Theatre. May 5, 1955. Gwen
Verdon (Lola), Stephen Douglass (Joe), Ray Walston (Applegate). Principal
songs: "Whatever Lola Wants," "Heart." 1,019 performances.

The Music Man. Music, book, and lyrics by Meredith Willson. Majestic Theatre. December 19, 1957. Robert Preston (Harold Hill), Barbara Cook (Marion Paroo). Principal songs: "Seventy-six Trombones," "Goodnight, My
Someone," "Lida Rose," "Trouble." 1,375 performances.

Gypsy. Music by Jule Styne; book by Arthur Laurents; lyrics by Stephen
Sondheim. Broadway Theatre. May 21, 1959. Ethel Merman (Rose), Jack
Klugman (Herbie), Sandra Church (Louise). Principal songs: "You'll Never
Get Away From Me," "Everything's Coming Up Roses," "Together," "Rose's
Turn." 702 performances.

Fiorello! Music by Jerry Bock; book by Jerome Weidman and George Abbott;
lyrics by Sheldon Harnick. Broadhurst Theatre. November 23, 1959. Tom
Bosley (Fiorello La Guardia), Patricia Wilson (Marie), Howard da Silva (Ben).
Principal songs: "Till Tomorrow," "Little Tin Box." 795 performances.

Bye Bye Birdie. Music by Charles Strouse; book by Michael Stewart; lyrics by
Lee Adams. Martin Beck Theatre. April 14, 1960. Dick Van Dyke (Albert
Peterson), Chita Rivera (Rose Grant), Susan Watson (Kim MacAfee), Dick
Gautier (Conrad Birdie), Paul Lynde (Mr. MacAfee), Kay Medford (Mrs. Pe-
terson). Principal songs: "Baby, Talk To Me," "A Lot of Livin' To Do," "Put
On A Happy Face." 607 performances.

How To Succeed in Business Without Really Trying. Music and lyrics by Frank
Loesser; book by Abe Burrows, Jack Weinstock, and Willie Gilbert. 46th St.
Theatre. October 14, 1961. Robert Morse (J. Pierrepont Finch), Rudy Vallee
(J. B. Biggley), Charles Nelson Reilly (Bud Frump). Principal songs: "I Be-
lieve In You," "Grand Old Ivy." 1,417 performances.

A Funny Thing Happened on the Way to the Forum. Music and lyrics by Stephen Sondheim; book by Burt Shevelove and Larry Gelbart. Alvin Theatre. May 8, 1962. Zero Mostel (Pseudolus), Brian Davies (Hero), Preshy Marker (Philia), Ron Holgate (Miles Gloriosus), David Burns (Senex), Jack Gilford (Hysterium). Principal songs: "Lovely," "Free," "Everybody Ought To Have A Maid." 967 performances.

Hello, Dolly! Music and lyrics by Jerry Herman; book by Michael Stewart. St. James Theatre. January 16, 1964. Carol Channing (Dolly), David Burns (Horace Vandergelder), Eileen Brennan (Irene Molloy), Charles Nelson Reilly (Cornelius), Jerry Dodge (Barnaby). Principal songs: "Hello, Dolly!," "It Only Takes A Moment." 2,844 performances.

Hair. Music by Galt MacDermot; book and lyrics by Galt MacDermot and James Rado. Public Theatre. October 29, 1967. Transferred to the Biltmore Theatre on April 29, 1968. James Rado (Claude), Gerome Ragni (Berger), Shelley Plimpton (Crissy). Principal songs: "Aquarius," "Starshine." 1,836 performances.

Your Own Thing. Music and lyrics by Hal Hester and Danny Apolinar; book by Donald Driver. Orpheum Theatre. January 13, 1968. Rusty Thacker (Sebastian), Leland Palmer (Viola). Principal songs: "Baby! Baby!," "Do Your Own Thing." 933 performances.

Company. Music and lyrics by Stephen Sondheim; book by George Furth. Alvin Theatre. April 26, 1970. Dean Jones (Robert), Elaine Stritch (Joanne). Principal songs: "The Ladies Who Lunch," "Someone Is Waiting," "Side By Side By Side." 706 performances.

Follies. Music and lyrics by Stephen Sondheim; book by James Goldman. Winter Garden Theatre. April 4, 1971. Alexis Smith (Phyllis Stone), John McMartin (Ben Stone), Dorothy Collins (Sally Plummer), Gene Nelson (Buddy Plummer). Principal songs: "Broadway Baby," "I'm Still Here," "Losing My Mind." 522 performances.

Grease. Music, book, and lyrics by Jim Jacobs and Warren Casey. Eden Theatre. February 14, 1972. Barry Bostwick (Danny Zuko), Adrienne Barbeau (Betty Rizzo), Carole Demas (Sandy Dumbrowski). Principal songs: "Look At Me, I'm Sandra Dee," "Greased Lightnin'." 3,388 performances.

The Wiz. Music and lyrics by Charlie Smalls; book by William F. Brown. Majestic Theatre. January 5, 1975. Stephanie Mills (Dorothy), Dee Dee Bridgewater (Glinda), Hinton Battle (Scarecrow), Tiger Haynes (Tinman),

Ted Ross (Lion). Principal songs: "Ease On Down The Road," "Believe In Yourself," "Who Do You Think You Are?" 1,672 performances.

A *Chorus Line*. Music by Marvin Hamlisch; book by James Kirkwood and Nicholas Dante; lyrics by Edward Kleban. Public Theatre. April 15, 1975. Transferred to the Shubert Theatre on July 25, 1975. Robert LuPone (Zach), Donna McKechnie (Cassie). Principal songs: "One," "What I Did For Love." Still running in April 1982.

Annie. Music by Charles Strouse; book by Thomas Meehan; lyrics by Martin Charnin. Alvin Theatre. April 21, 1977. Andrea McArdle (Annie), Reid Shelton (Daddy Warbucks), Dorothy Loudon (Miss Hannigan). Principal songs: "Tomorrow," "You're Never Really Dressed Without A Smile." Still running in April 1982.

Dreamgirls. Music by Henry Krieger; book and lyrics by Tom Eyen. Imperial Theatre. December 20, 1981. Jennifer Holliday (Effie Melody White), Sheryl Lee Ralph (Deena Jones), Ben Harney (Curtis Taylor, Jr.), Cleavant Derricks (James Thunder Early). Principal songs: "And I Am Telling You I Am Not Going," "Cadillac Car," "Love Love You, Baby." Still running in April 1982.

Appendix 2

Adonis
by William Gill

A D O N I S

A Disrespectful Perversion of
"Pygmalion and Galatea," "Marble Heart" and Common Sense
by William Gill

Music By Beethoven—Audran—Suppe—Sir Arthur Sullivan—Planquette—
Offenbach—Mozart—Hayden—Dave Braham—Eller—and many more too
vastly numerous to individualize, particularize or plagiarize.

CHARACTERS

ADONIS (Statue—Poet—Artist—Political Economist—Athlete—Aesthete—
Astronomer—Bicyclist—Dancer—Singer—Elocutionist—Postmaster—Dry-
goods Clerk—Tonsorial Artist—Drug Clerk—Old Clothes Man—General Hero and High Kicker)

TALAMEA   (A Sculptress who, like most of her sex, is in love with her own
creation)

ARTEA   (A Goddess—Patroness of the Fine Arts. N.B. The student will
vainly search for this character in the Heathen—or any other—mythology.
She was invented to suit the fell purpose of the Author)

DUCHESS OF AREA   (Aesthetic to the verge of eccentricity, rich to the
    verge of millionarism, sentimental to the verge of gush)

ROSETTA (A simple village maiden—The happy possessor of a clear con-
    science and a soprano voice)

BUNION TURKE   (Her father. An unblushing appropriator of the stock-in-
    trade of one of the characters made famous by the Madison Square The-
    atre Management)

MARQUIS DE BACCARAT   (A highly polished villain. This character will
    be recognized by the nude eye of all lovers of the modern society play)

LADY PATTIE ⎫
    "   MATTIE ⎬   The daughters of the Duchess—
    "   HATTIE ⎪   Professional Beauties
    "   NATTIE ⎭

GYLES—NYLES—STYLES—BYLES   (Ordinary everyday Rustics)
UNGBO—TUNGBO—BUNGBO—SHUNGBO   (Oriental Menials)
JENKYNS—SNOOKS—JEAMES—BUTTONS   (Caucasian Menials)

THE BRIC-A-BRAC GUARDS

TROOPS—TIGERS—VILLAGERS

THE GOOD OLD-TIME CHORUS

ADONIS

—Act 1—

Sc. 1   *Studio of Talamea the Sculptress.*
          Articles of statuary &c, scattered about.
          Curtains over alcoves.
          Talamea discovered C. dreaming.
          Opening chorus invisible.

    "We winged messengers that watch o'er mortals in their dreams,
    "Are guarding thee fair votaress, whose hopes have been deferred,
    "We fain would grant thy wishes—bid thee hope to keep—
    "Thrice blessed one thy prayer is heard."

Tal.   (*after chorus*) Come Artea, come to thy pupil and worshipper—Come,
        oh come.
        (*Artea comes thro' trap C.—Music*)
Art.   I am here Talamea—why this hasty summons?

Tal.	Great Goddess—I am miserable.
Art.	You the great sculptress miserable, wherefore?
Tal.	I love.
Art.	Your art? Yes!
Tal.	Of course I do—But it is not that.
Art.	What then? A man?
Tal.	No alas!
Art.	A lass! You love a lass?
Tal.	(*Indignantly*) A lass! The idea!
Art.	Not a lass—not a man—what then? A Politician?
Tal.	A Statue!
Art.	A mere figure!
Tal.	Don't speak of its figure in that tone of voice. Its figure is perfect—glorious—divine—
Art.	Bartholdi's Statue of Liberty enlightening New York Harbour?
Tal.	By the time *that* is placed in position, I fear I shall be too ancient for any amatory passion.
Art.	Who then—or rather what then is the object upon which you waste your affections?
Tal.	Waist? Ah yes, it has a beautiful waist—I should know, for I shaped it—it is my masterpiece—Adonis!
Art.	In love with your own statue?
Tal.	It is not mine.
Art.	You sculped it.
Tal.	And sold it to the Duchess of Area, before it was finished, and now that I adore it I feel disinclined to fulfil my part of the bargain.
Art.	The Duchess can compel you! She has the Law and Roscoe Conkling on her side.
Tal.	What care I for Law and Roscoe, so long as I have you on mine.
Art.	How will my friendship benefit you?
Tal.	Great Goddess! By your marvellous power you can accomplish miracles—perform one now on my behalf.
Art.	What miracle?
Tal.	Endow with life my beautiful statue, and let him choose between his maker and his purchaser, and then he will be mine.
Art.	Be not too sure of that! If men of flesh and blood cannot be trusted by our sex, what warranty have we that a mere marble effigy of the perfidious male will be more conscientious?
Tal.	I am willing to abide the test—My ardent passion must kindle a corresponding spark in my loved one's bosom.
Art.	I will consider the matter. In the meanwhile let me behold this wonder, also the other works you have accomplished since my last visit.

(Talamea draws aside curtain of alcove R. 1.E and discloses figure, "The Plunge.")

Art.   What do you call that?

Tal.   A Society Belle—taking her first plunge into the whirlpool of Fashion.

(Draws aside Curtain R. 2.C. and discloses facsimile of woman washing boy at tub, "You dirty boy.")

Art.   That!

Tal.   Dame Columbia washing her dirty boy "Politics" in Terra Cotta.

Art.   Terra Cotta! You mean soap—at least so it appears—The old lady has her work cut out.

(Talamea draws curtain of alcove L. 1.E. and discloses figure of "Justice"—patch on eye &c.)

Art.   Justice has evidently been in the wars.

Tal.   Wishing to be as correct as possible, and noticing that Justice nearly always gets a black eye in our courts—I have so depicted her.

(Draws curtain L. 2.E. and discloses the "Three Graces")

Tal.   Behold the far-famed Three Graces.

Art.   A charming group of pretty faces.

Tal.   And now for Adonis.

(Draws curtain C. and discloses "Adonis")

Art.   A tinted statue? Charming idea. Did it take my Talamea long?

Tal.   Yes. No time Talamea's *tinted.* Learn of its accomplishments. By means of reeds and pipes operated upon by a key, I can make my statue sing. *(Turns crank in pedestal—statue sings—bus.)*

Art.   Wonderful! *(Draws curtains)*

Tal.   Ought not one so accomplished to be alive?

Art.   Some people with souls not attuned to music might say no, but I will consider the matter, and if a crisis in your affairs demands action on my part, I will do what I consider best for your happiness.

(Exit R.U.E. Enter Ladies Pattie, Mattie, Hattie and Nattie L.U.E.)
Chorus

"We are the Duchess daughters and behave as nobles ought,
"We've come to see the statue, our gracious Ma has bought,
"They say it is a darling—with form that takes the cake,
"So an investigation we will proceed to make."

Mat.   *(Draws Curtain 1)* Is it this?

All.   No!

Hat.   *(Draws Curtain 2)* Is it this?

All.   No!

Pat.	(*Draws Curtain 3*) Is it this?
All.	No!
Nat.	(*Draws Curtain 4*) P'raps it's this?
All.	No! no! no! no!

"No not this, no not this,
"Where then is the beauty?
"No not this, no not this,
"Then seek it out we must.
"No not this, no not this,
"To find it is our duty,
"Of figures we have heard it is the topmost, the topmost upper upper crust."

Pat.	Stay here is an alcove that we have not yet explored. (*Draws curtain & discloses Adonis—Chord!*)
Mat.	Oh how beautiful.
Hat.	Isn't it lovely.
Nat.	What a shame it's only marble.
Pat.	What grace in that nostril.
Mat.	What symmetry in that eyebrow.
Hat.	What indications of strength in those biceps.
Nat.	And what lovely calves.
All.	Oh for shame.
Pat.	(*Drawing Curtain to*) That chaste conception of the sculptor's art shall not be subject to such an unladylike remark.
Nat.	I don't care—They are lovely—The difference between us is, that *I* say what *you* think, whilst you *think*, but don't *say* it. (*Duchess appears outside*)
Mat.	Hush, here's Ma. (*Enter Duchess followed by the Marquis de Baccarat L.U.E.*)
Duch.	How now girls—you here?
Pat.	Yes Ma. We heard you say you were coming to claim your statue, and we wished to participate in your enjoyment.
Duch.	I have brought the Marquis de Baccarat with me. (*Salute*) He cannot believe that my statue is as beautiful as it has been represented.
Marq.	Nay your grace, you misunderstood me. What I said was that I could find no beauty in anything when you were by—You absorb it all.
Duch.	Flatterer! (*Goes R. with daughters*)
Marq.	I am a marquis, and I mix in the very highest circles of society, but they little dream that I am a polished villain, and have adorned all the very modern melodramas and society plays in that character— When I smile I always show my teeth, and usually perform all my

atrocities enveloped in a dress suit, repressed passion, patent leather shoes, and a buttonhole bouquet. But beneath the surface (*raises vest and discloses pistols and daggers*) Ha! Ha!

(*Enter Talamea and Artea L. 1.E.*)

Art.   (*Aside to Talamea*) Unseen by them, I will watch over you, and do what I think is for the best.

Tal.   Thanks. (*To Duchess*) Welcome Madame to my Studio.

Duch.   Talamea, I have come to claim my statue. Is it finished?

Tal.   It is.

Duch.   Then I will take it with me. An express wagon is without.

Tal.   It shall not go!

All.   What?

Duch.   But I have paid you for it.

Tal.   Here is your money. (*Flings purse*) Take it back! I cannot part with my Adonis!

Duch.   This is very extraordinary conduct—But let me see the statue and then I will consider your proposition.

Tal.   (*Aside*) If she see it, she will insist. (*Aloud*) You cannot see it.

Marq.   Excuse me! The duchess not only can but shall! Behold! (*Opens Curtain C.*)

Duch.   Exquisite! What beauty is expressed in every line.

Tal.   (*To Artea*) She too is captivated—she will insist.

Pat.   Ma's enraptured.

Nat.   I should think so, who could resist those calves.

Girls.   For shame.

Duch.   Marquis! What do you think of it.

Marq.   Well I think the limbs are altogether out of proportion—the modelling of the left arm is atrocious—the head is ridiculously small—the treatment of the ear criminally bad—the size of the feet positively abnormal—but in other respects it's doing quite well.

Duch.   Nonsense. It is like life, and oh girls, it is mine.

Tal.   I will not part with it.

Duch.   I will invoke the law.

Tal.   (*Seizing mallet*) Rather than that I will dash it into fragments.

All.   Hold!

Art.   (*Aside to Tal*) I will grant your desire, give life to the Statue and it shall decide. (*Aside*) It is kill or cure.

Tal.   Madame, the statue shall decide to whom it will owe allegiance.

All.   The statue!

Marq.   Ridiculous!

Duch.   Monstrous!

Girls.   Pre-pos-ter-ous!

Tal.	True love has ere this moved hearts of stone to pity, why not this marble?
Duch.	Why should I accept the challenge when the statue is mine already.
Tal.	If you refuse, the statue ceases to exist.
Duch.	Well be it so—let Adonis choose between us.
Tal.	Adonis! I have no palace to offer thee as shrine, no wealth to lavish on thy beauty—nothing to bestow but love, as boundless as the starry heavens, as deep as the unfathomed sea. Choose!
Duch.	I am wealthy and noble—a Duchess in my own right, and the proud mother of four professional beauties, the sale of whose photographs alone brings me in a princely income—I can adorn you with jewels, stuff you with dainties, buy you a single eye-glass and a silk umbrella with a silver knob! Choose!

(*Artea waves wand—pedestal with statue descends stage—statue slowly comes to life & turns to Talamea—Duchess gives money & statue turns to Duchess*)

All.	It lives.

Concerted

Air: "Clog-Dance"

This marvellous awakening is really very singular
And scientists would say no doubt, it cannot be a fact.
It moves with such agility in truth it is quite wingular,
A miscellaneous lot of grace within that form is packed.
It flings its feet around about with very much rapidity,
Its pigeon wings and heel and toes, no one can e'er excel.
For marble which we all suppose is gifted with solidity
It does the light and airy steps most excellently well.

(*Adonis jumps from pedestal*)

It's alive, oh it's alive, of that there's not the slightest doubt,
It's alive, oh it's alive, see how it flings itself about,
It's alive, oh it's alive, some secret wire sure there must be,
Perhaps it's worked, Perhaps it's worked by electricitee.

Marq.	This a foe will be to me.
Tal.	Traitor from my presence flee.
Duch.	Is it love that tugs me so, or tight-lacing, I don't know.
Girls.	Lovers all away I cast, Mr. Right has come at last.
Adonis.	If some one don't hold down, red I soon shall paint the town.
All.	It's alive &c.

(*Closed in*)

Sc. 2	*Exterior of Turke's cottage.* (*Enter Rosetta from cottage.*)

Ros.   I am the village beauty and I weigh one hundred and twenty pounds—
       in addition to which I am very poor. I am pursued by all the lordly
       vilyuns for miles around, and my father is a poor old miller who
       threatens to turn me out of his doors if I listen to the voice of the
       noble charmer, but never fear I shall preserve my honor even to the
       verge of emotional insanity and shall continue to sing my country
       ditties as long as the public will stand them. (*Song and exit R.*)
       (*Enter Marquis L.*)

Marq. Ah yonder trips Rosetta the village beauty—to help her poor old fa-
       ther who has fallen into a mud hole—how tenderly she helps the old
       man out of the ditch, with what grace she wipes the wet clay from
       his nickel-plated hair—she shall be mine. If all else fail, I will stran-
       gle the old man and fly with her to my chateau in Alaska. But first I
       must wed the Duchess of Area—her lands and mine adjoin, and will
       make a noble property—should she refuse—(*raises vest significantly*)
       Ha! Ha! The polished villain will get in his fine work.
       (*Enter Duchess L.*)

Duch. Ah marquis, you here.

Marq. I cannot deny it—I am—and you, if you will permit me to remind
       you, are here too.

Duch. Strange as it may appear, I am.

Marq. Singular coincidence.

Duch. You will excuse me Marquis, I have a call to make.

Marq. On the old miller?

Duch. Yes—Rosetta his daughter tells me the old man is failing fast, so
       knowing his delicate appetite, I have brought him a cheap edition of
       Talmage's Sermons and a pair of leather shoe laces.

Marq. Ever charitable—but the old miller is not in—in other words he is
       out.

Duch. Then I will return home

Marq. Will you permit me to accompany you?

Duch. Thanks—but—

Marq. Duchess, for 22 years I have pursued your hand in marriage, during
       which time you invoked the law to aid you, you promised to be
       mine, each time you went back on your vows and married the other
       fellow. Now Duchess, I claim your promise, *I* want to be the other
       fellow this time.

Duch. Never!

Marq. Perhaps if the fascinating Adonis were to crave you, you would not
       be so obdurate.

Duch. He is at least a gentleman!

Marq. A block of marble a gentleman—a fellow with no family whatever—

Why he hasn't even a father and mother. Perhaps even now some of his connections may be acting as Philadelphia doorsteps.

Duch. Never allude to this again, Marquis, or I shall be exceedingly angry. (*Going—stops and hold out hand*) We shall still be friends?

Marq. Friends! Always!

Duch. So kind of you. Shall we see you at the ball this evening?

Marq. Well I should hop—I mean I should hope so.

Duch. Until then good afternoon. (*Exit*)

Marq. Good afternoon—That is how the truly polished villain disguises his emotions, but beneath the lips that smile he wears the teeth that rend—I'll be revenged! I don't know how just at present, but my time will come later on. The practised dramatist never fails to give the polished villain a chance—and when my chance comes—(*bus*) Ah here comes Rosetta—first to deceive her in the orthodox way, and then for the Duchess.
(*Enter Rosetta R.*)

Ros. My instincts inform me that this is another lordly villain! Will they never cease to persecute an innocent and hungry maiden?

Marq. Fair Rosetta, be mine, and my love and my wealth shall make you the envied of all your village companions.

Ros. (*Aside*) Now to bestow a stinging rebuke upon him. (*To Marq.*) My lord, you ought to be ashamed of yourself.

Marq. Such bitter sarcasm does not befit those lovely lips—Rather let me close them with a kiss that their ripeness and freshness invite.

Ros. My lord, if my lips were as fresh as you are, I should give up chewing gum, and take to ice as a steady diet.

Marq. Come fly with me to sunny Michigan and there—

Ros. Back base lord, and learn that the village maiden would rather vegetate on two thousand a month with an honest son of poverty than mingle in all the Hampton Court gaieties your wealth can procure.

Marq. I love to see you angry—it imparts a brighter lustre to your eyes. (*Seizes her*) Nay struggle not. I have a wrist of steel and a cheek of iron.
(*Enter Bunion Turke R.*)

Bun. Ah my child in the grasp of a lordly vilyun? I know he's a lord because he wears a diamond stud—and I'm sure he's a vilyun for no honest man could afford such clean hands.

Ros. Oh Father!

Bun. Go 'way—I sent you out to get the morning paper off my neighbor's doorstep, and instead of doing so you linger here and permit a lordly vilyun to embrace you.

Ros. I didn't permit—I couldn't help myself.

Bun.   No matter. I'm a worthy old miller and it's a part of my daily routine to turn my daughter out of doors—No better opportunity than this may occur, and I'm not the man to throw a chance away.

Ros.   Father, I implore!

Bun.   Go! You are no longer a child of mine! I close my poor but honest door upon your form as I close my poor but honest heart upon your memory. Go! Starve! Die! I care not! Stand back, I am lost to thee forever. (*Goes into house & returns*) If you don't go for that newspaper in less than no time—there'll be a row in this precinct. You hear my poor but honest voice. (*Exit*)

Marq.   His wild words make even me shudder—Is he often taken thus?

Ros.   Yes, bless him—about four times a week. He lives in hope that his conduct will reach the ears of some East End manager who will engage him as second old man.

Marq.   Why they wouldn't let him tend stage-door.

Ros.   My lord I will not stay to be insulted.
                    (*Trio*)

Ros.   Go basest lord from me away
        I do not want your love.
        And to your serpent I'll not play
        The part of stricken dove.

Marq.   Proud maiden tho' you scorn me thus
        I love you all the more.
        (*Enter Bunion*)

Turke.   Here come I say just stop this fuss
        In front of my honest door.
        (*Repeat "Go basest lord"*)

Ros.   Oh Pa! Oh Pa, he's a naughty, naughty man.

Marq.   I love and I adore your daughter more than I can speak.

Bun.   Here come I say I'm a tough old citizen
        And if you worry me, I'll knock you into Sunday-week.
        Oh Pa, Oh Pa &c.
        (*Exit Marq: L. Ros. & Bun: R.*)
                End of Scene

Sc. 3   *Garden*
        Enter Guards—march and chorus
                Tic-a-tac, tic-a-tac, bric-a-brac, bric-a-brac,
                  All day long.
                Tic-a-tac, tic-a-tac, bric-a-brac, bric-a-brac,
                  Is our song.

(*Repeat*)
Yes we are a happy lot
Guards of honor to her grace.
We are always on the spot
Whenever danger menaces the place. (*Exit*)

(*Enter Talamea & Artea*)

Tal. See there he is! Oh how beautiful—yet oh! how false.

Art. Infatuated still?

Tal. And ever will be.

Art. He loves you not.

Tal. Do the birds and flowers love us? Yet who shall censure us for or-dering quail on toast, or buying "Jack" roses at 75¢ a bud?

Art. I should for one—for like the poor, are not the dudes always with us?

Tal. I do not love Adonis because I want to, but because I can't help myself.

Art. Then I will help you—I will transform him back into his marble state.

Tal. Then he will be mine again, and I can break him all up!

Art. Shall I?

Tal. No. No. Life to him is all joy, unimpaired by indigestion. What matters my misery, so long as he can enjoy a good square meal.

Art. You are very far gone.

Tal. With shame I confess it! My love for him robs me of all my rest, and my professional ambition—Look at my numerous orders—Here is a plaster cast of Jumbo, one of Jay Gould on a bust, as yet un-touched by chisel. (*Music*) But hark! Artea! What is that?

Art. It is the Duchess approaches, accompanied by her Tigers—let us retire, and watch Adonis go through his competitive examination.
(*Enter Boy-Tigers—Chorus "The Boys of the Boots-and-Breeches" from the opera of "The Merry Duchess." Then enter Guards, four daughters and Duchess*)

Girls. Welcome home Momma.

Duch. I hurried back from Monaco to see my Adonis, who is, you say, in so short a time so marvellously proficient.

Hat. He appears to gain his knowledge of history by intuition.

Mat. Were astronomers classified like actors—he would be a very star among them.

Nat. If Barnum were to see him he would sell his white elephant and engage Adonis as his prize card.

Pat. He sings like Sims Reeves, and soon will dance like an Alhambra ballet.

Duch. Indeed! Then he must be much older than he looks.

Girls. Oh Ma, he's a darling.

Duch. My dear girls restrain your ardour. So much gush does not become young ladies. (*Aside*) Four pretty and marriageable girls may be in the way of a mother whose heart is still impressionable.

Pat. He comes.

Hat. How gracefully he bounds along.

Nat. He scarcely bends the tender cabbage leaves on which he steps.
(*Enter Adonis—song: paraphrase of the "Susceptible Chancellor" from "Iolanthe"*)

Duch. Where have you been, Adonis?

Adon. Here, there, and everywhere! Chasing sunbeams—catching measles, I should say weasels asleep—and diving for ducks in the frog-pond.

Duch. A perfect child of Nature.

Adon. Nature. I adore Nature—Especially human nature, don't I girls? (*Flirting*)

Duch. Leave Nature alone for the present and let me see your Art.

Adon. My 'eart? Thinks't thou I wear my 'eart upon my sleeve for Duchesses to peck at?

Duch. Your A.R.T. not your H.A.R.T.

Adon. Excuse me—H.*E*.A.R.T.

Duch. Wonderful—He already knows how to spell.

Adon. Yes, there's a spell about me that no one can resist.

Duch. Having *heard* of what you know will you permit me to *see* how you do it all?

Adon. All! Life is too short—a portion with pleasure—Having learned rapidly to imbibe—

All. Eh?

Adon. An education.

All. Oh!

Adon. Education you know is the system of finding out what is what.

All. Well what is what?

Adon. Oh well if you don't know what is what—I may as well go on to the next branch of study, which is lightning calculations upon the blackboard—Each one can get a clear insight into my system by watching the different expressions I throw into my left eye. Suppose for instance I were a census-taker asking the age of a young lady, and the young lady were—to say—the Duchess—not to say the Duchess wouldn't give her age (*aside*)—No one would take it—(*aloud*) I should approach the young lady in this way—Madam, what year were you born in (*Duchess whispers*) Quite a young thing! Now by putting down a row of figures on the blackboard in shorthand—adding them

	up, subtracting these, and dividing by 3, I find the Duchess is exactly 22 years, 4 months and 7 days old.
Duch.	Oh Adonis—Someone must have told you.
Adon.	Oh no. My system never fails. Now for Astronomy, all you have to do is to slip up on a piece of orange peel and you'll see all the stars the astronomers ever dreamt of.
Pat.	Arithmetic? If you hire a plumber at ten shillings a day for 20 days, what will he have at the end of the time.
Adon.	Your house and lot.
Nat.	Music!
Adon.	Ah music hath charms to soothe the savage beast—That's why they put a brass band round a dog's neck.
Nat.	In 3/4 time, how many beats are there in a bar?
Adon.	That depends upon the quality of the free lunch—And that concludes the first part of my entertainment.
Duch.	Refreshments here! I will serve the sandwiches myself—Will they not taste better from my hand?
Adon.	Oh for the touch of a sandwiched hand and the clink of the glass that we fill.
Duch.	What will you have, Adonis? Ham, beef or sausage.
Adon.	Oh my appetite like my nature is ethereal—Have you any moth's tongues on toast—or canary bird's wing—
Pat.	We're just out of them at present.
Adon.	Well, give me some hash.
Duch.	Serve the champagne (*bus*) Adonis, do you like champagne?
Adonis.	Well I should effervesce.
Duch.	May I offer you a glass.
Adon.	No thanks, I happened to drink about three bottles last night and I had a terrible dream—I dreamt that I was a celestial body, having been promoted from the position of "star" to that of a planet, and I was sailing along, when suddenly I perceived a speck coming towards me though the Milky Way. So I said to it: "From whence comest thou?" "From the earth," it made reply, "where I have been growing higher and higher and smaller and smaller every day."
All.	Why, what was it?
Adon.	The bottom of a strawberry basket. (*Enter Marquis*)
Duch.	What eloquence. Ah Marquis, is he not the modern edition of the admirable Crichton.
Marq.	Say rather the modern edition of a most insufferable snob.
Duch.	You are ungenerous.
Marq.	Pardon me, I am simply fly.

Duch.	He is highly accomplished.
Marq.	I will test him—H'm (*bowing*)
Adon.	That is not the way to bow sir, let me show you. (*Bus*)
Marq.	Insolent fellow.
Adon.	You are aggrieved? A bout at foils then, and the first touched apologizes.
	(*Tiger brings down foils*)
Marq.	One of these foils is shorter than the other.
Adon.	Then I will take the longer. En garde.
Marq.	Take care sir, I am the finest fencer in the Quartier Latin.
Adon.	And I my lord am the champion of the Rue de Pompadour.
	(*Bus: of fencing—Marquis is disarmed*)
Marq.	You fence exceedingly well sir.
Adon.	Your fence would never keep chickens out of a flower garden.
Marq.	We will resume this on some future occasion—Toujours prêt.
Adon.	Eh.
Marq.	Ne comprennez-vous pas le Français?
Adon.	What is that gibberish?
Marq.	I thought this highly accomplished gentleman spoke French.
Adon.	I do, but I didn't learn the accent in Soho.
Marq.	That is an excellent cigar you have there.
Adon.	Will you try one
Marq.	Thanks awfully.
Adon.	What do you think of it?
Marq.	It's like a man with a pat hand—it doesn't draw.
Adon.	It should be good—I paid two shillings for the two, buying them by the box.
Marq.	Indeed.
Adon.	Yes, one-and-tenpence for this.
	(*all laugh*)
Duch.	What exquisite readiness! Marquis, you must take a few lessons in repartee.
Marq.	It's your turn now sir, but we shall meet again.
Adon.	Not if I see you coming.
Marq.	I'll see you later.
Adon.	Not if I see your shadow.
Marq.	My time will come.
Adon.	Yes, I'll speak to Attenboro about it.
Marq.	Taunt on, sir, each insult I plant as a milestone on my road to vengeance. (*Aside*) Oh that I had him by the throat, I'd—but I keep forgetting that I am a polished villain—ranting is such very bad form. (*Aloud*) (*Bowing*) Ladies, your most obedient—Sir Ex-Statue, I salute you—But my time will come. (*Exit L.*)

Duch.	Now girls away—let my guards and servants vanish, I have a few words to say in private to Adonis.
	(*Exit guards and daughters*)
Duch.	Adonis, you must ere this have observed a tenderness in my manner towards you.
Adon.	All the daughters in love with me, and now the old girl! I'm solid with the whole family.
	(*Enter Rosetta R.U.E.—stands at top of stage*)
Duch.	Dearest it is because I have made up my mind to make you my fifth.
Adon.	Oh what a vision of bliss (*looking at Rosetta*)
Duch.	I am glad you feel that way about it. Then you do not think me positively ugly?
Adon.	If I could but catch the expression of her eyes, I'd have them set to music.
Duch.	Say you so dearest—then you may kiss me.
Adon.	Eh?
Duch.	I say you may kiss me.
Adon.	What have I done.
Duch.	I see dear bashful youth, kisses are as yet strangers to those beautiful lips.
Adon.	I'm willing they should remain strangers.
Duch.	Come I will teach you—Now I put my mouth so (*action*) and you do the same. (*Adonis grimaces*) No, not that way, so (*Adonis makes a mouth*) Now, one, two three.
	(*Rosetta comes between them*)
Ros.	Oh your grace.
Duch.	Bah! How provoking.
Adon.	Thank the Gods! I don't think I could have got through it.
Duch.	Girl, I went to your father's house this morning, on charitable thoughts intent—I heard he was starving, so I took him a pair of leather shoe laces—I withdraw my gift! Plead not! I am resolved—Adonis, don't stay out in the open air too long, it may rain, and you might get damp and get discolored, and I should never forgive myself. Au revoir mon cher ami. (*Exit Duchess*)
Ros.	Oh! What have I done that she should be so cross? (*She sees Adonis & is startled*) Oh how beautiful.
Adon.	Really—
Ros.	Excuse the abruptness of the remark but I couldn't help making it.
Adon.	Of course you couldn't—I'm not offended.
Ros.	How good of you to say so—you see I am a simple village maiden to whom beauty of your classic style is a staggerer.
Adon.	While your bucolic loveliness has stamped its trade mark upon my heart.

Ros.   You are as elegantly phraseological as you are physically majestic and we provincials are unaccustomed to such bewildering conjunctions.

Adon.   And your expressions smack more of the Bostonian elegantum than the Arkansasian vulgaris—Are you a product of the cornfields of Jayville?

Ros.   Yes, where were you born.

Adon.   I never was.

Ros.   Never was what.

Adon.   Never was born.

Ros.   Gracious.

Adon.   I was quarried.

Ros.   Quarried?

Adon.   Brought to the earth's surface by a gang of Italian laborers and a steam drill.

Ros.   What do I hear?

Adon.   In my earlier days I was to be pitied, for my youth was full of blasted hopes and I have been chiselled from my inception.

Ros.   That is the best thing you ever got off.

Adon.   Excuse me—the best thing I ever got off was a sixty-dollar overcoat. What I got *on* it—I decline to say.

Ros.   How old are you.

Adon.   That is hard to determine, as I belong to the stone age.

Ros.   You *astone-age* me. (*Crash*) Whatever's that?

Adon.   Somebody trying his voice.

Ros.   Oh! Pardon me but don't think me unmaidenly if I say that—I love you to distraction.

Adon.   Of course—you couldn't help yourself.

Ros.   Then may I hope.

Adon.   Hope on forever.

Ros.   Then I am yours.

Adon.   Then you and I are one. (*Duet & dance*)
(*Enter Bunion Turke*)

Bun.   Ah my daughter in the arms of a lordly vilyun!

Ros.   Father!

Bun.   Begone! You are no longer a child of mine—I close my poor but honest door upon your form, as I close my poor but honest heart upon your memory—Go! Starve! Die! I care not! (*Sees lunch on table*) Ah lunch. (*take off hat & sweeps lunch in it*)
(*Bunion is about to exit*)

Adon.   Stay, if you want lunch you must buy beer at the counter.

Bun.   Ah you would not only deprive the old man of his sole support, but you would deprive him of the consolation of a free lunch.

Adon.   Second old man, I love your daughter and would marry her, and would keep her upon all the luxuries that a railway-brakesman's salary can afford.

Bun.   If you will keep me—you can have her with pleasure.

Adon.   Keep you? I should think I would. Keep you in the woodshed and use you as a saw buck.

Bun.   Listen—Twenty years ago, upon a dark tempestuous night—when the thunder thundered, the rain rained, & the lightning lightened, seated at my little old log cabin in the lane—

Ros.   Oh Pa—you make me so fatigued.

Adon.   Is that old man your father?

Ros.   I cannot tell a lie—He is.

Adon.   (*tapping miller's chest—flour comes out*) What's that.

Bun.   That's emotion.

Ros.   He's a miller—But how are we to cross the river?

Adon.   (*turning upstage*) The Duchess's coal-barge approaches. (*Enter Artea & Talamea*)

Tal.   Hold!

Ros.   Who is that young party.

Adon.   A very old friend of mine.

Tal.   Oh Adonis, exquisite creation of my exuberant fancy, will you not pity me.

Art.   Not cured yet?

Ros.   (*to Adonis*) The lady knows you methinks.

Adon.   Yes, she used to be cashier at a restaurant that I used to go to.

Tal.   Inhuman monster!

Ros.   Don't you call my young man names. (*Raises hands theateningly*)

Adon.   Don't hit her with those, get an axe.

Tal.   Wed her, never!
     (*Enter Duchess*)

Duch.   Who talks of wedding.

Tal.   Adonis would marry this girl.

Duch.   Wed her, never! What ho! my guards (*enter guards &c.*) If he attempts to go, slay him on the spot.

Tal.   Oh do not handle him roughly, he may yet be brittle.

Art.   Fear not, I will save him.

Duch.   (*to Marq*) Marquis take that girl away. (*Marq. endeavours & gets buffetted*)

Bun.   Listen: Twenty years ago, when Rosetta was but a little prattler, I remember—(*Tiger diverts his attention with bottle*) Ah!

Art.   (*to Adonis*) Take this magic cane. It possesses a charm that will keep you from harm, and the knees of your pantaloons from bagging.

Duch.   Now let him be locked up & fed on bread and water till time and a reduced system of dieting shall have cured him of this mad infatuation.

Adon.   I will try the power of my magic cane.

(Finale: Chorus & picture—Ros. & Adon, up stairs in centre)

(Chorus)

He would away—here he must stay
Till he is gray—hidden away—
In tower grim, quick lock him in
Feed him on bread—straw for his bed.

Tal.   Oh spare him so beautiful
Altho' to me not dutiful.

Art.   Fear not I will protect him for your sake.

Ros.   I will go with him, so beautiful
And be to him so dutiful
And never cause him sorrow pain or ache.

Marq.   He shall not thus escape, for on his track I will be quick,
My vengeance he's excited, and to him I'll ever ever stick.

Adon.   Fools I do defy you, I'll be far far from your search
Unless this magic cane deceives me
And leaves me, and leaves me in the lurch.

Chorus.   Oh spare him, spare him Duchess
Let not his life be past in gloom
Oh spare him, spare him Duchess
And give him elbow room.
With head erect and martial step
He frowns upon his foe
His handsome face displays contempt
His nervous actions show.
That he intends, and quickly too
Away from here to press
And take his bride, the fair Rosette
On the *limitted* Express.

Picture—End of Act I

Act II

Sc. 1   *Rustic Sc.*

Chorus

Oh yes we are the chorus
The happy, happy chorus
That stand in lines upon the stage

In comic opera.
Stage-managers abhor us
And with rehearsals bore us
About the way we wear our hair
Are most particulair.
Our hands we use like this—
Like this we use our feet
And dance around so gracefully
To see us is a treat.
And sometimes flop like this:

Solo.	When at the forge I hammered horse-shoes
Male.	Opera, friends said, pined for me.
	Del Puente I quite shaded
	But in the chorus yet I be.
Chorus.	When at &c.
Solo.	When I gave up my board and wash tub
Female.	Upon the stage to tra-la-le
	I thought that wealth my voice would bring me
	But in the chorus yet I be.
Solo.	When I left Whiteley's ribbon counter
Male.	Upon the stage my high chest C
	I thought 'twould fill the world with wonder
	But in the chorus yet I be.
Cho.	When we left our occupations
	Just to seek variety
	We little thought as years rolled onward
	In the chorus yet we be.
	Oh yes we are &c.

(*Exit Chorus—Enter Tal. & Artea*)

Tal. Is it possible that Adonis has chosen this humble village for his abode?

Art. Speak not so slightingly of it—It is the county seat—and its gentlemanly treasurer has just absconded with the funds.

Tal. Then it is more progressive than I thought. But tell me, will Adonis never be mine again

Art. Yes but before that time he must be punished for his ingratitude, and from now on misfortunes shall crowd thick and fast upon him.

Tal. He'll not be hurt much, will he?

Art. The cane I gave him will protect him from bodily harm, but he shall find that in spreading the butter of pleasure too greedily upon the bread of life he is liable to be put off with the oleomargarine of disappointment.

Tal.   He'll never eat bread & butter when he can get cake.

Art.   In the meantime the Duchess will sue him for breach of promise—Rosetta deceived by the Marquis, will desert Adonis—and the daughters of your enemy disguised as Opera Bouffers will hither come and torture him with the old old melodies we all know so well and they sing so badly.

Tal.   Isn't the village set with more than ordinary care today?

Art.   In honor of the wedding of Adonis & Rosetta a village fete will soon take place.

Tal.   But they will not wed.

Art.   Fear not—here comes one whose villainous machinations will do much to prevent it.
         (*Exit Art. & Tal. R.—Enter Marquis L.*)

Marq. In this disguise I can defy detection. I have tracked Adonis to this village—I am going to serve him with the papers announcing the breach of promise instituted by the Duchess of Area—Then I will steal away his bride, the fair Rosetta—finally I will denounce him to the British Authorities as a dynamiter—Ah here comes Rosetta, I will hide in yonder grove and watch for an opportunity to accost her. (*Exit R.*)
         (*Enter Ros. L.*)

Ros.   All Nature smiles upon my wedding day—Everything smiles upon my wedding day—Even the horny handed day laborer goes into the village inn and at the bar he smiles—upon my wedding day. But where is Adonis—Ah I see him. Even on our wedding morn he is too conscientious to neglect his customers. He comes disguised as a milkman, with the pride of conscious manhood on his brow and a milk-pail in his hand. (*Exit L.*)
         (*Enter Adonis with milk-pail*)

Adon. (*going to pump and filling can*) Ah my dear old friend—even if my neighbour's cows go back on me—they will never miss the milk until the pump runs dry. I serve city customers, and if I gave them the real article I might poison them. Yes it may be bad water, but it never would be bad milk—Now tell me dearest what think you of my plan of disguise.

Ros.   Excellent Adonis—you're a genius.

Adon. Yes but I must get something for the wedding breakfast.

Ros.   Don't be long.

Adon. Fear not—I'm only going to the mountains to shoot a few Welsh rabbits for breakfast. (*Exit*)

Ros.   Dear thoughtful fellow—He knows that I adore Welsh rabbits—I wonder if he will kill any—oh yes—he's sure to kill something, if he doesn't know the gun's loaded.

(*Gun heard off—Adonis rushes on & falls in Rosetta's arms*)

Adon. I am hemmed in by my pursuers. On the winding road in the valley are the Duchess daughters, on the other side I beheld the Marquis whilst on the mountain's peak seven miles off I saw a troop of soldiers playing seven-up with Buffalo Bill's Indians—but fear not Rosetta, I will never be taken alive—Rather than that I would eat pies of your making and die of indigestion.

Ros. I have an idea.

Adon. Give it to me—I may write a play.

Ros. My wardrobe is extensive.

Adon. It would have to be.

Ros. You put on one of my dresses, and thus disguised you may elude them and when the Marquis & Duchess have gone—the wedding can proceed.

Adon. Brave girl—you have the great big head—I will do as you say—More would I say to you, if more I had to say, but as it is, farewell, farewell. (*Exit into cottage*)

Ros. Will he baffle them—he must for I have bought my wedding dress, and I'm going to be married or know the reason why.

(*Enter Marquis*)

Marq. Can you direct me to Earlswood.

Ros. Oh, a stranger.

Marq. No, I'm a New York detective working up a case against one Adonis Marble who has lately purchased a milk [*sic*] here.

Ros. Adonis—a detective—oh heaven. What is it? A bank robbery or a simple embezzlement.

Marq. He is connected with a gang of determined ticket speculators and he has two wives sitting in the lap of poverty in Paris, Kentucky.

Ros. Can you give me proofs of this.

Marq. I can, and if it doesn't come naturally I'll manufacture it.

Ros. Then I am yours.

Marq. I haven't asked for you yet.

Ros. I know that, but in all well constructed dramas whenever the villain gives the leading lady proof of her husband's baseness, she always says then I am yours.

Marq. Thanks awfully—but I'm rather busy just now—Meet me on your mountain's forehead—I mean brow—and there I will give you the proofs of your would-be husband's perfidy.

Ros. Stay, have I now seen those features before—are you not—

Marq.   No, I'm another fellow altogether. Besides it's impossible for you to recognize me, I have put on another suit of clothes, and I've changed my name—and if you did recognize me you would spoil all my plans, and you wouldn't do that—now would you.

Ros.   Not for the wide wide world.

Marq.   Don't mention it.

Ros.   Trust to me—the simple village maiden has not been to the theatre on a free pass for nothing.
          (*Enter Bunion L.*)

Bun.   Ah my child in the grasp of a lordly vilyun—go, you are no longer a child of mine.

Ros.   Oh Pa don't bother—I am busy now. (*Exit*)

Bun.   I never can have an agreeable recreation—no matter, I'll call her into the kitchen, & curse her over the stationary washtub—I'll show her. Oh before I go (*to Marq:*) Twenty years ago when I was eating peas with my knife—(*Marq. seizes him by the throat*)

Marq.   Another word and I'll ring the fire alarm.

Bun.   Well I'll tell that story if I die for it (*goes to pig*) Twenty years—(*pig falls—goes to donkey*) Twenty years—(*donkey rears, kicks Marquis— Bunion exits*)

Marq.   Ah here come the Duchess's daughters to the village fete—I mustn't stay here, but in the midst of the fete Adonis shall meet his fate in me—I'll serve him with the papers, steal away his bride, denounce him as a dynamiter & then the polished villain will be master of the situation. (*falls over pig*) 'Tis nothing, merely an everyday hog- currence—I'm glad I said that it's so villainous. (*Exit R. Enter Five Venuses—Five Olivettes—Five Crusoes and Five Merry Wars*)

Pat.   Now let each one do the best she can to win the heart of the man she loves.

Hat.   See a village girl approaches, you retire girls whilst I question her about Adonis. (*Exit chorus and enter Adonis from house L. dis- guised as girl*)

Adon.   Ah strangers!

Hat.   Don't be afraid pretty one, we won't hurt you.

Adon.   Please don't—I'm so timid you know.

Mat.   Sweet innocent.

Adon.   What are you? Drummers?

Mat.   No.

Adon.   I wish you were—I like drummers.

Nat.   Do you? Why?

Adon.   They are so cute! When they come to our village they always stay at Pa's hotel—and they are so kind.

Nat.	You must be glad of anything to break the monotony of a village life.
Adon.	Oh we have lots of fun here. Sometimes theatrical troupes come along—and they generally stay quite a while.
Pat.	They get so fond of the place?
Adon.	No—They find it hard to get their trunks out—I'm going on the stage, when I get older.
All.	You!
Adon.	Yes—One of the actor gentlemen heard me speak a piece, and he said that if Pa would receipt his board bill, he'd get me an engagement.
Nat.	What was the piece?
Adon.	It's real good—would you like to hear it?
All.	Very much.
Adon.	Well don't laugh at me and I'll speak it just as I did in the schoolroom.
	(*Recitation delivered in an awkward girlish manner*)
All.	Capital.
Nat.	Now can you tell us anything about Adonis Marble.
Adon.	Oh that must be the young man that was here yesterday. He was peddling clocks.
Nat.	Peddling clocks.
Adon.	Yes, and I learned how to dance from him.
Hat.	Can you dance too?
Adon.	Oh yes—we have lots of practice. We have a sociable here once every six months, and we finish in this manner. (*Dances*)
	(*After Dance exit Daughters & enter Marquis*)
Marq.	Ah! a village divinity.
Adon.	Now you just stop that.
Marq.	Oh I can't.
Adon.	Well stop that—(*strikes Marquis who falls—exit Adonis into house*)
Marq.	Ah well I suppose it serves me right—with the Duchess and Rosetta to bend my knee to as a suitor I ought to be satisfied: "Ne sutor ultra crepi—" (*Donkey kicks Marq.*) When all is quiet I'll return and poison that donkey.
	(*Enter Bunion & Villagers*)
Bun.	Now lads and lasses come on, the sports will be opened with a circus by Messrs. Tills, Sills, Bills & Gills.
	(*Circus bus*)
	(*At end of Circus, re-enter Adon. & Rosetta*)
Ros.	My friends, let me present to you my future husband.
Adon.	They are Opera Bouffers.

Ros.   Opera Bouffers, oh, how shocking.

Adon.   Now lads and lasses let's on to the refreshment counter.
            (*Enter Marquis*)

Marq.   Is Adonis Marble anywhere about?

Adon.   I am he.

Marq.   In the name of the commonwealth of Sag-Harbor, I serve you with
            these papers (*bus*) (*Marquis knocked down by bricks, barrel & club*)
            However, as you appear to be rather busy about here I'll call some
            other time. (*Marquis exits—in passing donkey is kicked*)

Adon.   I'll go & get the cake. (*enters house*)
            (*Re-enter Marquis*)

Ros.   (*to Marq.*) Now sir your proofs.

Marq: They are here—behold this copy of the daily paper—"We regret to
            inform our numerous readers that we have learned the painful news
            that Adonis Marble snores."

Ros.   He cannot be so base—I refuse to believe it.

Marq: Come with me.
            (*Enter Adonis & Recitative*)

Adon.   Where is Rosetta—I have here the cake.

Cho.   He takes the cake.

Marq.   False one she is here

Cho.   Yes she's all there.

Ros.   I've learned the painful fact you snore.

Cho.   Oh horrid fact he snores.

Adon.   I do not snore

Cho.   No.

Adon.   I cannot snore.

Cho.   What?

Adon.   For on my nose I wear a clothespin.

Cho.   He wears a clothespin.

Ros.   Produce the clothespin and I am yours.

Adon.   Oh Heavens!

Cho.   What?

Adon.   I've lost the clothespin.

Bun.   My boy your magic cane.

Cho.   Oh joy he's safe.
            (*Bus. &*)
                        (*Closed-in*)

Sc. 2   (*Front Scene—Landscape*)
            (*Enter Talamea & Artea*)

Art.	Cheer up Talamea, your trials are nearly over. This very day I will restore Adonis to you.
Tal.	Gracious Patroness—How can I think you & show my gratitude.
Art.	By holding fast to your art. Genius & Love have never yet, and never can agree—both are autocrats and cannot brook a rival.
Tal.	But tell me what has become of Adonis since the day you rescued him from the power of the Marquis.
Art.	Guided by my unseen influence he wended his way to a village nearby and sought employment in a store—in one short week so talented a financier has he proved to be, that he has driven his employer into bankruptcy, and been given the post of receiver by the creditors.
Tal.	How clever he is.
Art.	In that obscure spot he hopes to elude the vengeance of the Marquis & the consequences of his rash flirtations.
Tal.	Did he not madly love Rosetta?
Art.	No. In conferring life upon him I studiously refrained from meddling with his heart, that still is marble & he will ere long pray to be restored to the inactivity of his former existence.
Tal.	Oh haste that consummation so devoutly to be wished. (*Exeunt*) (*Enter Marquis*)
Marq.	I have frequently observed that my time would come—it has—a week has elapsed since the day on which Adonis eluded my vengeance and although the soldiers have been on his track ever since, he is still at liberty and I am still disfigured—But no matter, I have alienated from him the affections of his village beauty whom I have secretly married—I'll destroy the wedding certificate—shoot the clergyman who married us—then I'll wed the Duchess—kill Adonis—poison the donkey & I think by that time I shall be the most polished villain of this and every other age.
Ros.	(*who has overheard Marq.*) Indeed—
Marq.	You here—then you've heard—
Ros.	All. A very neat plan that of yours my lord Marquis, but you have neglected to include my say in the combination.
Marq.	My dear—
Ros.	You aspire to be the greatest villain of the age. Why you couldn't earn your living as a confidence gentleman.
Marq.	Rosetta!
Ros.	I am but a simple village maiden but I know how to get a divorce, so beware.
Marq.	You really must have misunderstood what I said.
Ros.	Then don't misunderstand what I say—I shall leave South Lambeth and reside in Paris. I shall entertain the nobility—I will have my

portrait painted by Meissonier, and tear it up, if he doesn't make me twice as handsome as I am—and you as my husband, must foot the bill.

Marq.   What have I caught on to?

Ros.   A simple village maiden, who clad in her innate nobility of soul must ever be a match for the titled imbecility of an effete despotism.

Marq.   Oh spare me.

Ros.   Oh I can spare you at any time—but your wealth and title I will cling to, even as a prima donna holds to the D of her upper register.

Marq.   Oh lor!

Ros.   I am a simple little village maiden, but I know my rights. (*Exit*)
                              (*Change of Scene*)

Sc. 3   (*Country Store Set*)
        (*Adonis discovered*)

Adon.   Having been appointed receiver of this bankrupt stock I can safely say that in two weeks I shall be sole proprietor—in the meantime I have several disguises here by which I can escape the vengeance of the Marquis.
        (*Enter woman with basket for eggs*)
        Eggs—yes ma'am—How will you have them—raw, fried or omelette. Raw? One dozen raw.
        (*Enter young lady for dry-goods*)
        Why Miss Doolittle, how do you do. Ribbons? Yes miss—That was a delightful time we had at the party last week was it not? This shade? Yes—one or two yards? Bertie Spoondyke and I had quite a time after we left there—oh yes—Bertie drank so much ginger in his lemonade that he got real brave—Wanted to waylay a policeman and ask him what time it was—wasn't he awful? Bertie's real strong too—he carries his cane four blocks without dropping it—Lace tidies? Yes, 36¢ apiece—yes, it is rather dear for tidies. Well, we'll make it two for a half—Is there anything else? That'll be 36 and 18—that'll be a dollar five—good morning—now mind you come to our next sociable—we're going to have real oysters in the stew—good morning. And so I suit my style to the articles I dispense.
        (*Enter soldier*)

Soldier. Is this the drug store—I want a little Vichy please—and a little whiskey. (*fills glass*)

Adon.   25¢ please.

Soldier. I'll take a little more please. (*again fills glass*)

Adon.   8¢ please.

Soldier.	Why how's that?
Adon.	Wholesale price. Ah, a customer at the—(*Exit soldier*) postoffice (*changes to Dutch postmaster*)
Girl.	Any letters for Phoebe Perkins?
Adon.	How do you spell Phoebe, with an F or a fee—no letters for Phoebe with a fee. A postmaster should be a happy lot—He is a man of letters, a card well posted, and has all the stamps at his command. (*Enter Marq.*)
Marq.	Any letters for the Marquis de Baccarat?
Adon.	The Marquis! (*disappears*)
Marq.	That voice! Can it be possible that under that exterior lie hidden the classical outlines of Adonis—if it should be he—I have the troops outside and—
Adon.	D'ye vant to buy a coat?
Marq.	Not at present—where's the postmaster?
Adon.	He's gone to get reappointed.
Marq.	I'll call again.
Adon.	Better buy a suit of mourning for yer oder eye.
Marq.	Not this evening.
Adon.	Good morning.
Marq.	I'm convinced 'twas he, now to call in the troops. (*Exit*)
Adon.	Discovered but not lost, of course not, a man can't be lost if he is discovered. (*changes rapidly to barber*)
Marq.	Hi! You sir.
Adon.	You're next! Shave or haircut?
Marq.	Where is the postmaster. I want to see him.
Adon.	Well go and see him.
Marq.	But where has he gone to?
Adon.	Seein' you round I guess he's gone to lock up his money.
Marq.	(*to soldiers*) He can't have gone out there without our seeing him. You search the house whilst I watch on the outside—ah here comes that old man—he'll be wanting to tell me a story about—
Bun.	(*outside*) Twenty years ago—
	(*Exit Marquis. Enter Bunion*)
	Why that was the Marquis—I wonder what could have excited my son-in-law.
Adon.	His son-in-law—Then Rosetta is false—so much the better—one complication the less. Next—
Bun.	Ah I want a shave—a quick shave—I'm in a great hurry—I have to catch the next train.
Adon.	Well you bet yer sweet life you'll catch it—I'm be de lightnin' tonsorial artist of the East Side.

Bun. Can you shave a man in ten minutes?

Adon. Why I killed a man last week in less time than that.
(*Bunion in chair*)

Bun. Eh?

Adon. Well now you'll catch dat train, now don't you fret—(*going on with barber's preliminaries*) Say ain't you makin' a mistake about de time dat train goes.

Bun. No I'm not and if you don't hurry up I'll miss—(*Adonis puts brush in his mouth*)

Adon. Don't you know better dan to speak wid your mouth open in a barber's chair.

Bun. Oh make haste.

Adon. Don't you worry—you'll catch that train. (*Barber bus—Bell on telephone rings—Adonis goes to it*) Hello! Eh? Oh yes! Say is dat de restaurant? What have you got for dinner? What? Didn't I pay dat last check? Oh! Dat's all right! Send me a glass of beer and a sandwich.

Bun. Come on! Do hurry up.

Adon. (*resuming work*) I know my business. (*Telephone—Adonis goes to it—flap falls—he takes out plate of sandwiches and beer*) Say! I'll match you whether I have my lunch first & shave you afterwards, or shave you afterwards and have my lunch first.

Bun. Oh do go on. (*Enter Woman with Basket*)

Adon. Good mornin' Mrs. Reilly, what kin I do for you—Two bars of soap? Yes Ma'am, ten cents. How's Lizzie?

Bun. (*fidgeting*) Come on.

Adon. Tell her I'll call to-night Mrs. Reilly wid de rest of de gang.

Bun. Can't you make haste.

Adon. What's de matter wid you? You'll catch dat train—de conductor's a friend of mine. (*Bus of shaving continued and finished*) What'll you have, bay rum or vitriol?
(*Enter Rosetta to drug counter*)

Ros. A glass of Vichy please.

Adon. Rosetta! (*Disappears*)

Ros. That voice! Adonis!

Bun. Rosetta!

Ros. Pa!

Bun. Begone you are no longer—

Ros. Oh rubbish.

Bun. Excuse me, I forgot the change in our circumstances. But when one gets into the habit of turning out his child every day for five years, it's hard to break the habit.

Ros.	Pa, I've seen him.
Bun.	I wish I could see him—look how he's left me.
Ros.	Adonis?
Bun.	No that confounded barber!
Ros.	'Twas he disguised.
Bun.	Adonis the barber—I'll go & tell the Marquis—
Ros.	The Marquis? I hate him—All my former love for Adonis has returned—I shall leave this miserable Marquis.
Bun.	Leave the Marquis! And deprive me of the yearly allowance he makes me to keep out of his sight? Go, you are no longer a child of mine. No don't go, Rosie—there's a secret in the family, a secret which at last you should know.
Ros.	A secret! What is it?
Bun.	Twenty years ago—
Ros.	Oh! Don't bother, I must be off to get a divorce before the court closes.
Bun.	But Rosie reflect.
Ros.	When I have procured a divorce then I will reflect. (*Exit.*)
Bun.	Go, you are—(*sees beer on counter*) Ah beer (*goes to drink it and it moves away on slide*) You are no longer a beer of mine. (*Exit*) (*Enter Duchess*)
Duch.	Not here, then I must have been misinformed. I have assumed this costume in order to captivate my Adonis. And if I cannot capture him by force of charms, I will by force of arms. I never will relinquish my search. (*Enter Marquis*)
Marq.	Ah Duchess, the very person I wished to see.
Duch.	You told me I should find Adonis here.
Marq.	Well he was here.
Duch.	Was me no wases, but where is he?
Marq.	I don't know—oh, but never mind him, let us attend to our own affairs. My wife has just gone out to get a divorce. She said she'd not be very long, and when she comes back, you and I will wed.
Duch.	Wed you!
Marq.	Yes awfully jolly.
Duch.	With such a face and figure! Never!
Marq.	But my dear Duchess—
Duch.	Never! Never! Never! (*Exit*)
Marq.	Adonis was here, and the chances are he'll come back again—let me see now what plan shall I decide upon (*sits in chair*) Let me consider—(*back of chair slips & throws Marquis over*)—a good idea— I'll hide here. (*Hides L.*)

(*Enter Adonis*)

Adon.   Now to make good my escape—first to secure the registered letters and then for the next train to Honolulu.
(*Marquis springs out in front of Adonis*)

Marq.   Trapped at last.

Adon.   Never! And when a man says never in a situation like this, nothing will satisfy the audience but blood.

Marq.   You shall be thoroughly satisfied. (*goes to telephone*)

Adon.   What would you do?

Marq.   Call out the troops. (*puts speaking tube to his mouth—flour comes out into his face*)

Adon.   (*with razor*) Beware of this razor, sir, I am razorlute.

Marq.   I care not, for I hone a razor too. Now we are equal.

Adon.   Not so—that razor was last used on a sewing-machine agent's cheek and has not been worth a cent since. (*They fight until Marquis leaps through mirror—change scene*)

Sc. 4   (*Front Scene—Fantastic Trees*)
(*Enter Artea & Talamea*)

Tal.   You promised to restore Adonis to me and now that he has made his escape, he is as far from me as ever.

Art.   He is safe in my sunny home, whither I have directed the footsteps of the Duchess and her retainers—I will take you there.
(*Scene changes to*)

Sc. 5   (*Home of the Genius of Art*)
(*Enter guards—march & form picture—Enter Duchess, Rosetta, Four daughters, Bunion & Marquis*)

Art.   Now Adonis shall choose for himself. Adonis!
(*Enter Adonis*)
Choose between this life of care and strife and the peaceful seclusion of your sculptress' studio.

Adon.   Oh take me away and petrify me—place me on my old familiar pedestal—and hang a placard round my neck:—"HANDS OFF."

Finale
(*Repeat portion of Finale of Act I during which Adonis is transformed into a statue, and ascends stage* [?]).
Curtain

Index

[231]